Review of Army Research Laboratory Programs for Historically Black Colleges and Universities and Minority Institutions

Committee on Review of Army Research Laboratory Programs
for Historically Black Colleges and Universities and Minority Institutions

Army Research Laboratory Technical Assessment Board

Laboratory Assessments Board

Division on Engineering and Physical Sciences

NATIONAL RESEARCH COUNCIL
OF THE NATIONAL ACADEMIES

THE NATIONAL ACADEMIES PRESS
Washington, D.C.
www.nap.edu

THE NATIONAL ACADEMIES PRESS 500 Fifth Street, NW Washington, DC 20001

NOTICE: The project that is the subject of this report was approved by the Governing Board of the National Research Council, whose members are drawn from the councils of the National Academy of Sciences, the National Academy of Engineering, and the Institute of Medicine. The members of the authoring board responsible for the report were chosen for their special competences and with regard for appropriate balance.

This study was supported by Contract No. W911QX-09-C-0054 between the National Academy of Sciences and the Army Research Laboratory. Any opinions, findings, conclusions, or recommendations expressed in this publication are those of the authors and do not necessarily reflect the views of the agency that provided support for the project.

International Standard Book Number-13: 978-0-309-31311-7
International Standard Book Number-10: 0-309-31311-2

Copies of this report are available from

Laboratory Assessments Board
Division on Engineering and Physical Sciences
National Research Council
500 Fifth Street, NW
Washington, DC 20001

Additional copies of this report are available from the National Academies Press, 500 Fifth Street, NW, Keck 360, Washington, DC 20001; (800) 624-6242 or (202) 334-3313; Internet http://www.nap.edu.

Printed in the United States of America

THE NATIONAL ACADEMIES

Advisers to the Nation on Science, Engineering, and Medicine

The **National Academy of Sciences** is a private, nonprofit, self-perpetuating society of distinguished scholars engaged in scientific and engineering research, dedicated to the furtherance of science and technology and to their use for the general welfare. Upon the authority of the charter granted to it by the Congress in 1863, the Academy has a mandate that requires it to advise the federal government on scientific and technical matters. Dr. Ralph J. Cicerone is president of the National Academy of Sciences.

The **National Academy of Engineering** was established in 1964, under the charter of the National Academy of Sciences, as a parallel organization of outstanding engineers. It is autonomous in its administration and in the selection of its members, sharing with the National Academy of Sciences the responsibility for advising the federal government. The National Academy of Engineering also sponsors engineering programs aimed at meeting national needs, encourages education and research, and recognizes the superior achievements of engineers. Dr. C. D. Mote, Jr., is president of the National Academy of Engineering.

The **Institute of Medicine** was established in 1970 by the National Academy of Sciences to secure the services of eminent members of appropriate professions in the examination of policy matters pertaining to the health of the public. The Institute acts under the responsibility given to the National Academy of Sciences by its congressional charter to be an adviser to the federal government and, upon its own initiative, to identify issues of medical care, research, and education. Dr. Victor J. Dzau is president of the Institute of Medicine.

The **National Research Council** was organized by the National Academy of Sciences in 1916 to associate the broad community of science and technology with the Academy's purposes of furthering knowledge and advising the federal government. Functioning in accordance with general policies determined by the Academy, the Council has become the principal operating agency of both the National Academy of Sciences and the National Academy of Engineering in providing services to the government, the public, and the scientific and engineering communities. The Council is administered jointly by both Academies and the Institute of Medicine. Dr. Ralph J. Cicerone and Dr. C. D. Mote, Jr., are chair and vice chair, respectively, of the National Research Council.

www.national-academies.org

COMMITTEE ON REVIEW OF ARMY RESEARCH LABORATORY PROGRAMS FOR HISTORICALLY BLACK COLLEGES AND UNIVERSITIES AND MINORITY INSTITUTIONS

ARMY RESEARCH LABORATORY TECHNICAL ASSESSMENT BOARD

JENNIE S. HWANG, H-Technologies Group, *Chair*
KENNETH R. BOFF, Socio-Technical Sciences
EPHRAHIM GARCIA, Cornell University
GEORGE T. GRAY III, Los Alamos National Laboratory
PRABHAT HAJELA, Rensselaer Polytechnic Institute

Staff

EVA LABRE, Administrative Coordinator
JAMES P. McGEE, Director
ARUL MOZHI, Senior Program Officer
ANDREA SHELTON, Administrative Assistant

LABORATORY ASSESSMENTS BOARD

Acknowledgment of Reviewers

This report has been reviewed in draft form by individuals chosen for their diverse perspectives and technical expertise, in accordance with procedures approved by the National Research Council's Report Review Committee. The purpose of this independent review is to provide candid and critical comments that will assist the institution in making its published report as sound as possible and to ensure that the report meets institutional standards for objectivity, evidence, and responsiveness to the study charge. The review comments and draft manuscript remain confidential to protect the integrity of the deliberative process. We wish to thank the following individuals for their review of this report:

Joseph S. Francisco, Purdue University,
Antoine M. Garibaldi, University of Detroit Mercy,
Carmelita Lamb, University of Mary,
Cato T. Laurencin, University of Connecticut,
William A. Lester, Jr., University of California, Berkeley,
John W. Lyons, National Defense University,
Eduardo J. Padron, Miami-Dade College,
Winston E. Thompson, Morehouse School of Medicine,
Antonia M. Villarruel, University of Michigan School of Nursing, and
Robert A. Winn, University of Illinois College of Medicine.

Although the reviewers listed above have provided many constructive comments and suggestions, they were not asked to endorse the conclusions or recommendations, nor did they see the final draft of the report before its release. The review of this report was overseen by Warren M. Washington, National Center for Atmospheric Research. Appointed by the National Research Council, he was responsible for making certain that an independent examination of this report was carried out in accordance with institutional procedures and that all review comments were carefully considered. Responsibility for the final content of this report rests entirely with the authoring board and the institution.

Contents

APPENDIXES

Summary

At the request of the Army Research Laboratory (ARL), the National Research Council (NRC) formed the Committee on Review of Army Research Laboratory Programs for Historically Black Colleges and Universities and Minority Institutions. The committee's work was guided by the following statement of task:

> An ad hoc committee to be named the Committee on Review of Army Research Laboratory Programs for Historically Black Colleges and Universities and Minority Institutions (HBCUs/MIs), to be overseen by the National Research Council's Laboratory Assessments Board, will be appointed to examine the ways in which HBCUs/MIs have used the ARL funds to enhance the science, technology, engineering, and mathematics (STEM) programs at their institutions over the past decade. The committee will also consider which elements among the ARL HBCU/MI programs reflect practices that are effective for assisting HBCUs/MIs in enhancing the STEM programs at the institutions and that could be considered by other Department of Defense agencies for application to their programs. The study will not include examination of: (1) ways in which the HBCUs/MIs have contributed to effective outcomes of ARL projects, nor (2) career developments of students after they complete participation in the programs. The committee will prepare a report that summarizes the findings of its review.

The committee did not review HBCU/MI programs that are funded by the Office of the Secretary of Defense (OSD) but administered by ARL. The committee was not asked to, and did not, review workforce-related aspects of ARL investments in HBCUs/MIs, although workforce-related considerations could be a future goal of ARL's HBCU/MI programs.

To address its charge, the committee gathered data in three primary ways: (1) discussions with representatives of national organizations whose members include minority institutions; (2) discussions with representatives from organizations within the Army, Navy, Air Force, and Office of the Secretary of Defense that sponsor programs supporting minority institutions; and (3) visits with administrators, faculty, and students at selected institutions that have received funding from the ARL under its HBCU/MI support programs. In addition, the members of the committee examined literature relevant to the study and shared information and experiences from their deep knowledge base and backgrounds relevant to the study.

The committee was charged to examine the ways in which HBCUs/MIs have used the ARL funds to enhance the STEM programs at their institutions and to consider which elements of the ARL HBCU/MI programs reflect practices that are effective for assisting HBCUs/MIs in enhancing the STEM programs at the institutions. The examination therefore focused on the practices whereby ARL administers its support to HBCU/MI institutions and the practices by which those institutions secure and apply that support to enhance their STEM programs. ARL provided to the committee detailed descriptions of available programs and processes whereby they are administered, as well as detailed quantitative data describing the recipients of support under each of its programs, including the funding level in each year for each institution and investigator supported. The HBCU/MI institutions visited by the committee were asked to describe their processes for securing support, administering programs and projects, and using the support to enhance their STEM programs. These institutions provided informative discussions pertinent to these issues, and they often supported their descriptions with anecdotes indicating their perceptions of successes and challenges with respect to STEM enhancement and ARL processes.

The institutions did not provide the committee with detailed quantitative data, including data on the career outcomes of the principal investigators supported in their programs, that could be subjected to analysis, and no such quantitative data relating to ARL programs for these institutions were available in the literature.

TOWARD A MORE DIVERSE, EFFECTIVE, AND EFFICIENT U.S. SCIENTIFIC WORKFORCE

The primary goal of Army Research Laboratory programs to fund and support research is to improve the combat readiness, efficiency, and survivability of U.S. warfighters by developing and deploying technologies that will serve that basic imperative. How do ARL programs that support investigators and programs at HBCU/MIs further the fundamental goals of improving U.S. warfighting capabilities? In the current technologically innovative era, the answer is straightforward: The United States must have a strong and expanding intellectually talented and practically trained workforce in science, technology, engineering, and mathematics (STEM) disciplines. Underrepresented minorities (URMs) account for a growing share of the overall U.S. population, and even though historically black colleges and universities and colleges and universities that serve other minorities are relatively few in number in the overall universe of U.S. colleges and universities, they continue to enroll a disproportionate share of minority students. These institutions are critical to the education and scientific and technical training of the minority engineers, mathematicians, and scientists on which the military depends for effective warfighting technologies. HBCU/MI universities are one of the ways the United States, including the Army and other military departments, can ensure a fully mobilized, diverse workforce.

In turn, the recruitment and effective education of intellectually talented students requires strong, dynamic academic institutions. The capabilities of colleges and universities matter as much as the credentials of the students they enroll. Effective science and technology education at HCBUs/MIs depends on the capacity of these institutions to attract and retain capable faculty, gifted undergraduate and graduate students, and postdoctoral fellows, providing them with appropriate facilities and infrastructure to support their scientific activities. ARL has contributed to building up the human and infrastructural capacities of HBCUs/MIs in the past, and the committee has looked for ways to enhance ARL's program impact on institution building in the future, confident that more capable HBCUs/MIs will, in turn, help America as a whole develop a more diverse and intellectually capable STEM workforce.

ARL has used its very limited—and currently declining—financial resources to have a positive impact on HBCU/MI institutional capabilities. The ARL annual budget from Army sources and from work it does for others is approximately $1.3 billion, of which approximately $700 million is Army research and development funding; the rest may be generally characterized as reimbursable work for other agencies. The average annual ARL support for all HBCU/MI programs has been approximately $11 million annually over fiscal years FY2011 through FY2013.

ARL grants and programs have helped to jump-start and build technical infrastructure and have encouraged the enrichment of STEM curricular offerings and graduate student training opportunities. ARL programs have also helped to raise faculty and student morale, provided invaluable opportunity for productive scholarship, and increased the number of intellectually talented STEM graduates. The U.S. Army needs an enlarged STEM-based workforce, and even with limited resources ARL can do a great deal to meet that overriding need by devising and deploying more effective grants and support for research and development at HBCUs/MIs. *ARL programs in support of HBCUs/MIs will prove most effective if they place their primary emphasis on enhancing institutional capacities for STEM disciplines in those institutions.*

Overall, the ARL programs supporting HBCUs/MIs are strong, well run, and commendable. These programs have, over many years, provided support to many HBCU/MI individual researchers and institutions, administered through a variety of programs accessible to HBCU/MI researchers and institutions and reported by recipients to be helpful in their development and expansion of STEM programs. Within the narrow confines of the committee's tasking—that is, examining institutional STEM improvements at HBCUs/MIs as a result of ARL funding—the existing strong program can be made even stronger.

ARL funding levels to engage and support STEM capabilities at HBCUs/MIs are not substantial in amount, relative to the size of ARL total annual funding or in absolute terms, and are declining. It has become increasingly necessary, therefore, for ARL to consider carefully its strategies for allocating funds to and across these institutions and to regularly and systematically assess the impact of its support on the successful development, maintenance, and growth of the STEM programs at these institutions. *However, neither the Army nor, specifically, ARL, has put in place written directives or a strategic plan for supporting HBCUs/MIs or for assessing the impacts of that support. A successful strategy will include, for example, a reasoned, balanced, and effective allocation of support for single principle investigator research and longer-term collaborative programs that encourage institution building, interactions with other institutions and funding agencies, and support of students through completion of degrees.* Involving HBCUs/MIs in collaborative programs cannot be effective unless ARL applies a proactive management to ensure that HBCUs/MIs are provided with meaningful and sufficient levels of tasking and funding within those programs. A more effective support program will also include systematic mentorship whereby ARL educates HBCUs/MIs as needed with respect to the processes of proposal development, project and program implementation, and administration. To be most effective, mentoring would extend beyond ARL, but with ARL support, so that successful institutions, HBCU/MI and non-HBCU/MI, help to educate fledgling HBCUs/MIs in the STEM institution-building process. Of course, it is important for HBCUs/MIs to take advantage of opportunities, not only for funding, but also for learning the collaborative and institution-building processes as they aspire to the levels of success demonstrated by other HBCUs/MIs and other high-performing research institutions.

It is disconcerting that the majority of principal investigators supported by ARL HBCU/MI programs have not been underrepresented minorities (URMs). Over the past 10 years, the numbers and percentages of the 220 principal investigators funded by ARL HBCU/MI programs have been: 102 (46 percent) Caucasians, 14 (6 percent) African Americans, 11 (5 percent) Hispanics, 47 (21 percent) Middle Eastern and South Asians, 46 (21 percent) East Asians, and 0 Native Americans.

There is a dearth of URM researchers at HBCUs/MIs in STEM fields relevant to ARL activities. Over the past 20 years, many URM faculty have left HBCUs/MIs and joined nonminority institutions, and there is a tendency for recent URM graduates with Ph.D.'s in STEM fields to join the faculty at nonminority institutions. Given the constraints on available URM faculty at HBCUs/MIs, ARL's funding of non-URM researchers at these institutions has helped to build institutional STEM capabilities at the institutions.

However, one indicator of the success of ARL programs supporting HBCUs/MIs will be an increase in URM STEM researchers, including those who receive ARL support for research that contributes to ARL's mission to contribute to the development of technologies that enhance the safety and effectiveness of Army warfighters—a primary requirement for ARL funding. To achieve this goal, it is first necessary that ARL make every effort to assure that ARL opportunities are widely known to HBCU/MI researchers, including those who have not participated in ARL-funded programs. It will also be necessary for ARL to support the students and postdoctoral researchers who will grow the ranks of future minority faculty and to help the institutions attract and retain the minority faculty. These concerns are among those reflected in the recommendations presented in this report.

The recommendations that follow are presented in the spirit of helping to make a strong ARL program supporting HBCUs/MIs even stronger and of helping commendably vibrant HBCUs/MIs continue to build their institutions' STEM capabilities. Additional recommendations are also provided in the final chapter of the report:

KEY RECOMMENDATIONS

1. ARL should create and disseminate a policy directive regarding its commitment and priorities for a credible and sustainable HBCU/MI support program. Specifically, ARL should articulate a vision and write a strategy to enhance STEM capability within its HBCU/MI-supported community and develop metrics to measure STEM capability improvement; metrics should include progress toward independence, including expanded funding relationships with other funding agencies.

2. ARL should examine the funding of collaborative projects involving HBCUs/MIs and non-HBCUs/MIs to ensure that the funding is equitable and that the tasking takes advantage of HBCU/MI capabilities. ARL should require HBCU/MI participants in ARL-funded collaborative, cooperative agreement projects to provide to ARL regular reports on their experiences with the project planning, execution, management, funding, and other collaborative interactions with the sponsoring ARL program manager and other participants in the collaboration team.

3. ARL should regularly assess which HBCU/MI activities have the most successful impact on the development, maintenance, and growth of their STEM programs and should rebalance funding according to those assessments.

4. ARL should consider NIH, NSF, and other URM funding incentive models in allocating support, from within its HBCU/MI funds, for URM U.S. citizen undergraduate and graduate research students, summer interns, postdoctoral fellows, and faculty researchers.

5. ARL should proactively engineer the participation of its HBCUs/MIs in multiyear cooperative agreements to ensure that there is adequate funding and time for those institutions to gain access to and procure equipment, support the completion of graduate and undergraduate student research, arrange for onsite or virtual internships with ARL laboratories and other laboratories, and develop the capacity to respond to redirection of funded research tasks by ARL program managers. As long as ARL continues its University-Affiliated Research Center (UARC) programs, it should regularly consider HBCUs/MIs for UARC designation or for formal partnerships with existing Army UARCs based on continual adequate technical performance and extant STEM talent and physical capabilities.

6. With HBCUs/MIs in mind, ARL should review its core and cooperative agreement Broad Agency Announcement (BAA) processes to minimize administrative burdens on university respondents and should, with input from HBCUs/MIs, consider mentoring opportunities to enable more awareness of and success in responding to BAA opportunities.

7. ARL should gather best practices from other agencies in order to design models of funding that systematically connect and integrate single principal investigator research at HBCUs/MIs, with the efforts carried out by multi-institutional research teams, to facilitate institution building and the development of entrepreneurial scientific leadership at the HBCUs/MIs.

8. The HBCUs/MIs should pursue more ARL-supported collaborative research funding. Led by faculty and institutional leadership, HBCUs/MIs should engage in research opportunities that include collaborative grants and contracts as well as single investigator research and development.

9. The HBCUs/MIs should continuously improve the efficiency and effectiveness of their offices of sponsored programs that assist their faculty in execution of ARL-supported research programs both on and off campus.

10. The HBCUs/MIs should expand their ARL-supported research by partnering with local industry and international sources. The HBCUs/MIs should build relationships with foundations wherein foundational resources are combined with ARL resources to extend the research portfolios of the HBCUs/MIs.

1

Introduction

At the request of the Army Research Laboratory (ARL), the National Research Council (NRC) formed the committee on Review of Army Research Laboratory Programs for Historically Black Colleges and Universities and Minority Institutions. This introductory chapter describes the statement of task that guided the work of the committee, the process whereby the committee gathered data and produced this report, and the organization of the report.

STATEMENT OF TASK

The committee's work was guided by the following statement of task:

> An ad hoc committee to be named the Committee on Review of Army Research Laboratory Programs for Historically Black Colleges and Universities and Minority Institutions (HBCUs/MIs), to be overseen by the National Research Council's Laboratory Assessments Board, will be appointed to examine the ways in which HBCUs/MIs have used the ARL funds to enhance the science, technology, engineering, and mathematics (STEM) programs at their institutions over the past decade. The committee will also consider which elements among the ARL HBCU/MI programs reflect practices that are effective for assisting HBCUs/MIs in enhancing the STEM programs at the institutions and that could be considered by other Department of Defense agencies for application to their programs. The study will not include examination of: (1) ways in which the HBCUs/MIs have contributed to effective outcomes of ARL projects, nor (2) career developments of students after they complete participation in the programs. The committee will prepare a report that summarizes the findings of its review.

The committee did not review HBCU/MI programs that are funded by the Office of the Secretary of Defense (OSD) but administered by ARL. The committee was not asked to, and did not, review workforce-related aspects of ARL investments in HBCUs/MIs, although workforce-related considerations could be a future goal of ARL's HBCU/MI programs.

The term "minority institution" includes academic institutions such as the following: historically black colleges and universities (HBCUs), Hispanic-serving institutions (HSIs), tribal colleges and universities (TCUs), and Asian-American and Native American Pacific Islander serving institutions (AANAPISI).

HBCUs are colleges and universities founded before 1964; they were originally intended to provide higher education to African American communities. Although their enrollments are becoming more diverse, the vast majority of HBCUs continue to be predominantly black institutions. HSIs are institutions that receive federal discretionary funding to improve and expand their capacity to serve Hispanic and low-income students. At these colleges and universities, undergraduate students that identify as Hispanic make up at least 25 percent of total enrollment. TCUs are colleges and universities associated with American Indian and Native Alaskan tribes. The federal government provides grants and related assistance to TCUs to enable such institutions to improve and expand their capacity to serve American Indian and Native Alaskan students. AANAPISIs are institutions that receive federal discretionary funding to improve and expand their capacity to serve Asian-Americans and Native American Pacific Islanders and low-income students. At these colleges and universities, undergraduate students who identify as Asian-American or Native American Pacific Islander make up at least 10 percent of total enrollment.

The set of minority institutions that have received support from ARL, and which were considered by the committee, included those that serve black, Hispanic, and Native American student populations. Throughout this report, the term HBCU/MI is used when these institutions are collectively referred to.

DATA GATHERING AND REPORT PREPARATION

The committee gathered data in three main ways: (1) discussions with representatives of national organizations whose members include minority institutions; (2) discussions with representatives from organizations within the Army, Navy, Air Force, and Office of the Secretary of Defense that sponsor programs supporting minority institutions; and (3) visits with administrators, faculty, and students at selected institutions that have received funding from the ARL under its HBCU/MI support programs. In addition, the members of the committee examined literature relevant to the study and shared information and experiences from their deep knowledge base and backgrounds relevant to the study.

The committee was charged to examine the ways in which HBCUs/MIs have used the ARL funds to enhance the STEM programs at their institutions and to consider which elements among the ARL HBCU/MI programs reflect practices that are effective for assisting HBCUs/MIs in enhancing the STEM programs at the institutions. The examination therefore focused on the practices whereby ARL administers its support to HBCU/MI institutions and the practices by which those institutions secure and apply that support to enhance their STEM programs. ARL provided to the committee detailed descriptions of the available programs and processes whereby they are administered, as well as detailed quantitative data describing the recipients of support under each of its programs, including the funding level in each year for each institution and investigator supported. The HBCU/MI institutions visited by the committee were asked to describe their processes for securing support, administering programs and projects, and using the support to enhance their STEM programs. These institutions provided informative discussions pertinent to these issues, and they often supported their descriptions with anecdotes indicating their perceptions of successes and challenges with respect to STEM enhancement and ARL processes. The institutions did not provide the committee with detailed quantitative data that could be subjected to analysis, and no such quantitative data relating to ARL programs for these institutions were available in the literature. The committee engaged in discussions with representatives of the following national organizations: Hispanic Association of Colleges and Universities (HACU), American Indian Higher Education Consortium (AIHEC), and the Council of Graduate Schools (CGS).

The committee engaged in discussions with representatives of the following Department of Defense organizations: Army Research Laboratory Outreach Program Office; Army Research Office/ARL Technology Integration and Outreach Division; Office of the Deputy Assistant Secretary of the Air Force for Science, Technology, and Engineering; Air Force Research Laboratory/Air Force Office of Scientific Research; Assistant Secretary of the Navy for Research, Development and Acquisition; Office of Naval Research; and Naval Surface Warfare Center, Naval Sea Systems Command.

The committee visited and engaged in discussions with representatives of the following academic institutions: City University of New York, Hampton University, Howard University, Morgan State University, Navajo Technical University, North Carolina A&T State University, Prairie View A&M University, and the University of New Mexico at Albuquerque. A teleconference discussion was conducted with administrators and faculty at the University of Puerto Rico at Mayaguez.

The committee also engaged in discussions with staff at the House Armed Services Committee, Subcommittee on Intelligence, Emerging Threats and Capabilities.

This report represents the committee's consensus findings and recommendations, developed through careful deliberations during four committee meetings and reviewed by a separately appointed Report Review Committee appointed by the NRC.

ORGANIZATION OF THE REPORT

This chapter discusses the statement of task that guided the committee's work and the methods of data gathering employed toward developing this report. Chapter 2 discusses the context within which the ARL programs operate. Chapter 3 discusses the general perspectives of the community of minority institutions, and Chapter 4 summarizes the impressions and findings garnered from visits to and discussions with selected minority institutions. Chapter 5 summarizes the committee's findings pertaining to ARL principal investigator and collaborative research programs for HBCUs/MIs. The committee's recommendations are presented in Chapter 6. The appendixes provide additional discussion of the impressions and findings garnered from visits to and discussions with selected minority institutions, a summary description of collaborative programs at ARL, biographical information on the committee members, and a list of acronyms found in the report.

2

Context of Army Research Laboratory Programs for Historically Black Colleges and Universities and Minority Institutions

The primary goal of Army Research Laboratory (ARL) programs that fund and support research is to improve the combat readiness, efficiency, and survivability of U.S. warfighters – by developing and deploying technologies that will serve that basic imperative. How do ARL programs that support investigators and programs at historically black colleges and universities and minority institutions (HBCUs/MIs) further the fundamental goals of improving America's warfighting capabilities? In the current technologically innovative era, the answer is straightforward: The United States must have a strong and expanding intellectually talented and practically trained workforce in STEM disciplines. Underrepresented minorities (URMs) make up growing parts of the overall U.S. population, and even though historically black and other minority-serving colleges and universities are relatively few in number in the overall universe of U.S. colleges and universities, they continue to enroll a disproportionate share of minority students.[1,2] These institutions are critical to the education and scientific and technical training of the minority component of the diverse cadre of engineers, mathematicians, and scientists on which the military depends for effective warfighting technologies. HBCU/MI universities are one of the ways the United States, including the U.S. Army and other military departments, can ensure a fully mobilized, diverse workforce.

The recruitment and effective education of intellectually talented students requires strong, dynamic academic institutions. The capabilities of colleges and universities matter as much as the credentials of the students they enroll. Effective science and technology education at HCBU/MIs depends upon these institutions' capacity to attract and retain capable faculty, gifted students, and postdoctoral researchers, providing them with appropriate facilities and infrastructure to support their scientific activities. *ARL has contributed to building up the human and infrastructural capacities of HBCUs/MIs in the past, and the committee has looked for ways to enhance the impact of the ARL program on institution-building in the future, confident that more capable black and minority-serving institutions will, in turn, help the United States develop a more diverse and intellectually capable STEM workforce.*

An active HBCU/MI program is under way at ARL. ARL identified for the committee five objectives for its programs supporting HBCUs/MIs:

1. Foster support for meritorious research proposals originating at HBCUs/MIs;
2. Assist HBCUs/MIs in strengthening their capability to conduct quality research of interest to the Army;
3. Assist in the development of or enhancement of science and engineering education programs at HBCUs/MIs;

---

[1] Dexter Mullins, "Historically black colleges in financial fight for their future," *Al Jazeera America*, October 22, 2013, http://america.aljazeera.com/articles/2013/10/22/historically-blackcollegesfightfortheirfuture.html.

[2] *Education Encyclopedia*, "Hispanic-Serving Colleges and Universities—The History of HSIs, HSIs and Latino Educational Attainment, Conclusion," http://education.stateuniversity.com/pages/2045/Hispanic-Serving-Colleges-Universities.html, accessed May 25, 2014.

4. Increase the participation of URMs in research and development; and
5. Coordinate ARL's HBCU/MI programs with similar programs in other federal agencies.

ARL has also noted as follows on its website:[3]

> The objective of the U.S. Army Research Laboratory's (ARL's) Historically Black Colleges and Universities/Minority Institutions (HBCU/MI) program is to address the projected shortfall of scientists and engineers among the diverse populations of the 21st century, leverage HBCU/MI technical capabilities to fulfill ARL requirements, and expand the involvement of HBCU/MIs in ongoing research at ARL. ARL presently has Education Partnerships with six HBCUs/MIs.

However, ARL reported that there is no formal policy directive at the ARL or the Army that describes a vision or strategy for the enhancement of STEM capability or other goals at HBCUs/MIs, and no metrics for assessing the success of ARL's programs that support these institutions.

ARL is charged with conducting research and development to support the technological needs of the Army and of others that support it with reimbursable funding. This charge is consonant with the objective of assisting HBCUs/MIs in strengthening their ability to conduct quality research of interest to the ARL's mission and is focused on technology results. ARL provides on its Internet website[4] the following description of the mission for its Army Research Office:

> The U.S. Army Research Laboratory's Army Research Office (ARO) mission is to serve as the Army's premier extramural basic research agency in the engineering, physical, information and life sciences; developing and exploiting innovative advances to insure the Nation's technological superiority.

To fulfill its charge to support the technology needs of the Army, ARL needs to consider its own workforce development and that of others who participate in advancement of Army technological capabilities. In that regard, fostering workforce diversity is an obvious goal; however, workforce diversity was not identified by ARL as one of the objectives listed above. Like most other organizations in the Department of Defense (DoD), ARL pursues admirable DoD-wide and national societal goals as recognized by the objectives of increasing the participation of underrepresented minorities in research and development, fostering support of meritorious research proposals originating at HBCUs/MIs, and helping HBCUs/MIs to develop and enhance STEM education programs.

ARL's stated objectives for its programs supporting HBCUs/MIs are not internally consistent and do not benefit from a top-level Army or ARL policy directive. ARL could establish a vision, strategy, and set of metrics to assess success in areas related to HBCUs/MIs: institutional STEM capability, contribution to Army STEM workforce diversity, contribution to Army technology advancement, and/or other DoD or national goals.

HBCU/MI INSTITUTION BUILDING

ARL has as one of its goals in funding HBCU/MI programs, enhancing the institutional development at the recipient institution. The highly respected array of research universities in the United States is a model that is now replicated in most developed and developing countries. The development of this educational system began soon after the Second World War, when few government science and technology (S&T) programs existed and few of them were carried out at universities. The few universities

---

[3] U.S. Army Research Laboratory, "HBCU/MI," last update/reviewed November 3, 2010, http://www.arl.army.mil/www/default.cfm?page=39.

[4] Ibid.

that offered advanced degrees in the sciences and engineering undertook modest amounts of research, often funded by local industry. Several schools had already begun to develop strong departments, based on the European models, and in many cases peopled by émigrés from the stressed European systems.

The success of government-funded research in the prosecution of that war (e.g., the atomic bomb, radar, and jet propulsion), highlighted in Vannevar Bush's report *Science, The Endless Frontier*,[5] led to the subsequent establishment of DoD agencies in the Air Force (AFOSR) and Army (ARO), mimicking the already existing Office of Naval Research (ONR), as well as organizations such as the Atomic Energy Commission, the National Science Foundation (NSF), the National Institutes of Health, and the National Aeronautics and Space Administration, each committed to funding research at universities. The university community responded; those universities that already had strong departments were in the vanguard, but there was funding and strong incentive for other universities to perform S&T research.

The typical university at the time was primarily a teaching institution, targeted at producing enough teachers, mining and petroleum engineers, and agriculture-related professionals to satisfy the local economy. The path to research was clear, however: identify an area of a discipline to pursue, hire appropriate faculty (especially proven senior researchers) to establish the base, obtain federal funding for the targeted area, recruit graduate students, acquire facilities, establish STEM curricula, and add additional faculty over time.

Federal research programs drove the selection of target areas for development and supplied the critical resources required to hire faculty, capitalize research facilities and equipment, and fund the graduate students and research projects. In this environment, through the 1980s, several of the small teaching colleges transformed into large research universities. Then the federal funding leveled off (in real terms) just as the U.S. population grew from approximately 238 million in 1985 to 317 million in 2014, with African Americans, Latinos, Native Hawaiian and other Pacific Islanders, and American Indian or Native Alaskans comprising at least 36.5 percent of the U.S. population by 2012.[6] Universities that achieved research success during the years of budget growth were no longer so readily or easily emulated. Areas for potential federal funding were focused on already targeted areas at research universities. It was in this new reality that HBCUs/MIs were making their way from primarily four-year teaching colleges to research institutions. Universities realized that true institutional development requires funding targeted areas of disciplines for many years as faculty and facilities are added and developed. The infrastructure requires continual funding to maintain and enhance program capability. When successful, HBCUs and similar latecomers to the university research scene pulled all of these elements together from disparate sources—for example, state money, local donors, foundations, research funding from various federal agencies, and industry partnerships—to form a STEM institutional capability.

Many HBCU/MI institutions receive grants for small, single principal investigator (PI) programs; some have been centers of excellence, others have participated in multiyear, multiperformer projects, and some have received funds for all three of these broad categories. It is the last two categories through which sizable and competitive STEM capabilities have been established. *All ARL HBCU/MI programs contribute to STEM programs, but those ARL programs that are multiyear and/or that collaborate with other research universities or with ARL researchers have achieved the most recognizable enhancement of institutional STEM capability*. Individual grants provide flexibility and (if so targeted) excellent start-up resources, but they are not very effective in developing significant local strength. Cooperative grants and contracts do offer such possibilities, but they too must be managed in a manner intended to accomplish a well-defined institution-building goal that yields mutual long-term value for the Army and the university.

---

[5] NSF Office of Scientific Research and Development, 1945, *Science, The Endless Frontier—A Report to the President,* United States Government Printing Office, Washington, D.C.

[6] U.S. Census Bureau, "State and County Quickfacts," http://quickfacts.census.gov/qfd/states/00000.html, accessed May 25, 2014.

STATUTORY CONTEXT

Starting in 1987 funding in support of HBCUs/MIs was governed by statute 10 USC 2323, wherein a goal was established that 5 percent of DoD contracts should go to small, disadvantaged businesses and HBCUs/MIs. In 2010, that statute was replaced by 10 USC 2362, which eliminated goals and "set asides" for small, disadvantaged businesses and HBCUs/MIs. Today, ARL, like other DoD organizations, continues to support HBCU/MI programs, but absent statutorily set goals.

OVERVIEW OF ARL FUNDING PATHS FOR HBCUs/MIs

ARL is the Army's corporate research laboratory, providing the underpinning science, technology, and analysis that enable full-spectrum operations. It has a substantial multidisciplinary in-house capability within seven divisions (six laboratory directorates and the ARO). It also issues contracts, grants, and cooperative agreements for work by extramural research performers (e.g., industry, other federal laboratories, and universities, including HBCUs/MIs).

The ARO invests Army basic research funding with extramural performers, principally universities. The ARO manages a core HBCU/MI program for ARL and the Army. The ARO also serves as the executive agent for the OSD's HBCU/MI funding (this funding is not considered in this report). Program managers in the other six divisions of ARL can and do include HBCUs/MIs in their programs. ARL's Outreach Program office monitors and encourages interaction with HBCUs/MIs.

ARL (including its ARO) selects HBCU/MI performers based on the quality of responses to Broad Agency Announcements (BAAs) or requests for proposals (RFPs). Grants or cooperative agreements to conduct research on particular technical matters are concluded between ARL and those HBCUs/MIs that are selected after reviewing the responses to BAAs. Grants are typically, but not exclusively, between ARL and a single PI and include some funding for his/her graduate or undergraduate research student(s). Cooperative agreements take several forms, but typically an HBCU/MI will be a partner with ARL, with another or other universities, or with industry. Contracts are concluded between ARL and a selected HBCU/MI performer when the applied research involves a deliverable product.

ARL has many options to deploy funding to HBCUs/MIs. The most common have been grants to single PIs, centers of excellence (which are being phased out), and multiyear, multiperformer cooperative agreements; Figure 2.1 depicts recent funding, and Chapter 5 and the appendixes describe the various programs in more detail.

While HBCU/MI programs are important to ARL for technical mission achievement, workforce enrichment, and DoD (or other national) societal priorities, the ARL investment in HBCUs/MIs is not large in relative terms and has even been declining in recent years. The ARL annual budget from Army sources and from work it does for others is approximately $1.3 billion, of which approximately $700 million is for Army research and development funding; the rest may be generally characterized as reimbursable work for other agencies. The average annual ARL support for all HBCU/MI programs has been approximately $11 million annually over fiscal years FY2011, FY2012, and FY2013. Figure 2.1 shows a top-level summary of the funding amounts and percentages of the total investment for the various ARL programs for HBCUs/MIs for 3 recent fiscal years. Multiyear, multiperformer programs in Figure 2.1 include Partnership in Research Transition (PIRT) programs, Cooperative Technology Alliances (CTAs), and Cooperative Research Alliances (CRAs).

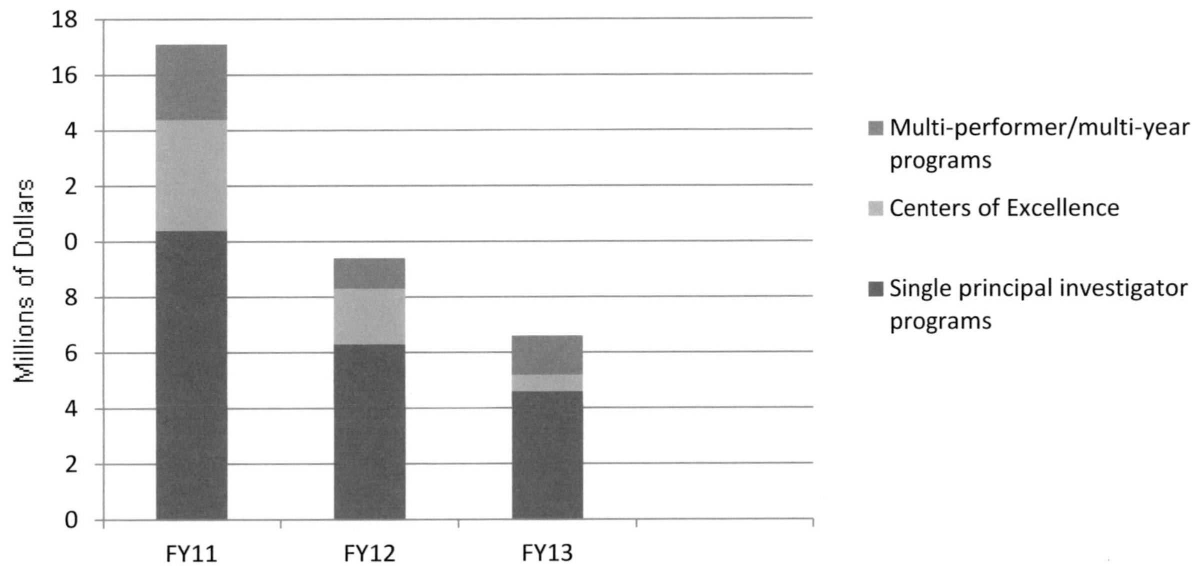

FIGURE 2.1 ARL investments in grants and programs to HBCUs/MIs, fiscal years FY2011-2013. Multiyear/multiperformer programs include the Partnership in Research Transition (PIRT) program, Cooperative Technology Alliances (CTAs), and Cooperative Research Alliances (CRAs). Single principal investigator programs include HBCU/MI ARO core grants in the following amounts: FY2011 ($3.6 million), FY2012 ($3.3 million), and FY2013 ($2.5 million).

ARL has supported a wide variety of HBCU/MI projects, with the stipulation that projects support ARL mission research, which covers such disciplines as computational science, materials science, ballistics, sensors, electron devices, survivability, vehicle structure and mobility, and human sciences. *While ARL HBCU/MI programs do contribute to STEM programs at HBCUs/MIs, they are not substantial enough to significantly enhance the institutional STEM capabilities. Moreover, they are relatively small when compared with ARL's total annual funding, and they are declining, especially for the HBCU/MI programs that have the most impact: those that involve multiperformer agreements and/or multiyear funding.*

HBCU/MI PROGRAMS AT OTHER MILITARY DEPARTMENTS

The committee was tasked to identify ARL practices with regard to HBCUs/MIs that may be useful to other military departments. The committee was briefed by the U.S. Air Force and the U.S. Navy on their respective HBCU/MI programs. The briefings and associated discussions described core HBCU/MI support in those military departments for single PIs and creative programs that focused on strategic HBCU/MI priorities, such as workforce development. *The discussions showed that the Army, Air Force, and Department of the Navy are intimately familiar with each other's arrays of HBCU/MI programs. Each military department has unique HBCU/MI programs tailored to its needs and culture. The various ARL multiyear, cooperative agreement programs are unique among the military departments, and they are well known by one another; considering their level of funding, discussions with selected HBCUs/MIs suggest that these programs successfully support institutional STEM improvement.*

SUPPORT FOR UNDERREPRESENTED MINORITY FACULTY AND STUDENTS

The majority of PIs supported by ARL HBCU/MI programs have not been African American, Hispanic, or Native American. Over the past 10 years, the numbers and percentages of the 220 principal investigators funded by ARL have been 102 (46 percent) Caucasians, 14 (6 percent) African Americans, 11 (5 percent) Hispanics, 47 (21 percent) Middle Eastern and South Asians, 46 (21 percent) East Asians, and 0 Native Americans.

These findings are disconcerting but understandable. A primary requirement for participation in ARL programs supporting HBCUs/MIs is that the supported STEM research contribute to developing and transitioning technologies that enhance the safety and effectiveness of Army warfighters. There is a dearth of African American, Hispanic, and Native American researchers at HBCUs/MIs in STEM fields relevant to ARL activities. Over the past 20 years, many minority faculty have left HBCUs/MIs and joined nonminority institutions, drawn by factors that include higher salaries and opportunities to work with more advanced equipment than that typically available at HBCUs/MIs. There is a tendency for recent URM graduates with Ph.D.'s in STEM fields to join the faculty at nonminority institutions, for similar reasons. Native American colleges and universities have particularly low levels of ARL-relevant researchers and supporting equipment.

Given the constraints on available URM faculty at HBCUs/MIs, ARL's funding of non-URM researchers at these institutions has helped to build institutional STEM capabilities at the institutions. The funded researchers establish and maintain STEM research programs, instruct undergraduate and graduate students, work collaboratively with postdoctoral researchers, and secure funding and equipment to support the students and postdoctoral researchers.

To achieve the goal of increasing the number of URM researchers working in STEM areas at HBCUs/MIs, it is first necessary to make every effort to reach out to HBCUs/MIs to make sure that ARL opportunities are widely known. Sponsoring broadly announced, periodic information dissemination symposia at which opportunities and program processes are elucidated would be helpful, especially if followed by proactive and regular mentoring of HBCU/MI candidates for funding on the BAA selection and program execution processes and on identification of other funding and collaboration opportunities within the Army.

It will also be necessary to support the students and postdoctoral researchers who will grow the ranks of future URM faculty and to help the institutions attract and retain the URM faculty. ARL could consider ways to expand the support at HBCUs/MIs of students serving as research assistants on ARL-sponsored projects. Multiyear (e.g., a 5-year norm) grants for single principal investigators and collaborative/cooperative agreements would support graduate students through completion of their theses and dissertations. This approach could be beneficially augmented by increasing the sponsorship of internships at ARL and by virtual internships with ARL laboratories and other laboratories. Other agencies (e.g., NIH and NSF) offer models for consideration that support minority undergraduate and graduate research students, summer interns, postdoctoral fellows, and faculty researchers.

To help attract and retain URM researchers at HBCUs/MIs, ARL could seek ways to provide public recognition and visibility for the sponsored research. Mentoring HBCU/MI faculty in the processes for applying for sponsored research and helping to alleviate administrative burdens associated with such research would contribute to overcoming current administrative challenges that could otherwise be one factor motivating researchers to move from HBCUs/MIs to larger institutions with more administrative support. Increasing HBCU/MI involvement in collaborative funded programs could be associated with opportunities for HBCU/MI investigators to share equipment with collaborators, and such collaboration also yields rewards with respect to research opportunities, recognition, and visibility in the S&T community.

Many research students at HBCUs/MIs are not from underrepresented minority (URM) groups, and some (especially graduate students) are not U.S. citizens.

The HBCU student body is still overwhelmingly black,[7] while Hispanic-serving institutions (HSIs) are increasingly highly heterogeneous.[8] Because of the relative homogeneity of the HBCU student population, the likelihood that URMs have access to cutting-edge research activities is higher at HBCUs than at HSIs.

There are 105 HBCUs in 20 states, the District of Columbia, and the U.S. Virgin Islands; though they represent less than 3 percent of all colleges and universities, they are responsible for awarding 18 percent of all degrees earned by black undergraduates. Although their enrollments are becoming more diverse, the vast majority of HBCUs continue to be predominantly black institutions: Black students make up more than 90 percent of the enrollments at 47 HBCUs and more than 75 percent of the enrollments at 80 HBCUs; moreover, black students are in the minority at only 7 HBCUs.

Most HSIs were not originally established to serve a particular student population. HSIs are generally characterized by their enrollment ratios rather than by their institutional mission, though there are several exceptions. HSIs represented 6 percent of all institutions of higher education in 2003, enrolling about half of all Latino undergraduates.[9] The number of HSIs has grown dramatically, but there are only unofficial lists like the Department of Education's list of High Hispanic Enrollment Institutions[10] or information contained in Title V Developing Hispanic-Serving Institutions Program (Historical List of all Grantees).[11] The Excelencia in Education group reported that the nation had 370 HSIs in 2013, an increase of roughly 60 percent from the 242 colleges that met the definition in 2003.[12]

Despite the fact that many ARL-supported faculty and research students at HBCUs/MIs have not been from URM groups, ARL HBCU/MI funding has generally enhanced STEM capability at the funded institutions and therefore has the potential to benefit URMs in STEM learning and research.

For at least its collaborative and multiyear HBCU/MI programs, ARL encourages summer internship programs, at ARL or similar Army facilities, for research students supported by ARL HBCU/MI funds. Such arrangements can be useful to the student and the funded research by, for example, exposing the student to leading-edge research efforts and researchers collaborating on ARL projects, providing access to sophisticated equipment at ARL, and providing familiarity with ARL staff, equipment, and facilities that may encourage subsequent applications for positions at ARL or later work with ARL in other contexts. However, such internships can conflict with progress-toward-degree plans or can be geographically inconvenient.

---

[7] M. Christopher Brown II and Ronyelle Bertrand Ricard, The honorable past and uncertain future of the nation's HBCUs, *National Education Association Higher Education Journal*, Fall 2007, pp. 117-130, http://www.nea.org/assets/img/PubThoughtAndAction/TAA_07_12.pdf.

[8] Lindsey E. Malcom-Piqueux and John Michael Lee, Jr., "Hispanic-Serving Institutions: Contributions and Challenges," College Board Advocacy and Policy Center Policy Brief, October 2011, http://advocacy.collegeboard.org/sites/default/files/11b_4853_HSBC_PolicyBrief_WEB_120110.pdf.

[9] *Education Encyclopedia*, "Hispanic-Serving Colleges and Universities—The History of HSIs, HSIs and Latino Educational Attainment, Conclusion," http://education.stateuniversity.com/pages/2045/Hispanic-Serving-Colleges-Universities.html, accessed May 25, 2014.

[10] U.S. Department of Education, "Accredited Postsecondary Minority Institutions," http://www2.ed.gov/about/offices/list/ocr/edlite-minorityinst-list-hisp-tab.html, accessed May 25, 2014.

[11] U.S. Department of Education, "Title V Developing Hispanic-Serving Institutions Program Historical List of All Grantees," http://www2.ed.gov/programs/idueshsi/hsi-allgrantees.pdf, accessed May 25, 2014.

[12] C. Dervarics, Hispanic-serving institutions continue growth with more poised to join ranks, *Diverse: Issues In Higher Education*, February 25, 2014, http://diverseeducation.com/article/60920/.

3

The Community of Historically Black Colleges and Universities and Minority Institutions

Challenges remain despite intense efforts of a multitude of institutions, over recent decades, to provide science, technology, engineering and mathematics (STEM) access to underrepresented groups. The legal background to these efforts is dynamic. The Supreme Court upheld Michigan's ban on the use of race as a factor in admissions to state universities, while in a recent ruling "the court ordered tighter scrutiny [in Texas] of race-conscious admissions, preserving the principle that affirmative action is permissible in some circumstances."[1]

Martin Luther King, Jr., observed that "life's most urgent question is: what are you doing for others?" and Nelson Mandela expressed the belief that "education is the most powerful weapon you can use to change the world." These statements succinctly define the fundamental role that HBCUs/MIs play and continue to play in fostering education through research in STEM programs. These institutions provide a critically valuable space where the promises of our democracy are accessible to citizens underrepresented in higher education in a nation experiencing dramatic shifts in demographics.[2] Education within HBCUs/MIs contributes directly to the expansion of a nationally diverse scientific leadership and workforce because it is carried out within models of research and education where access, excellence, and impact[3] are at the heart of their models of education. The long-term support and sustained funding provided by ARL and other federal programs to these institutions has been and continues to be critical to their ability to contribute to the nation's workforce capacity in STEM.

Recent reports by the National Academy of Sciences[4,5] and by the Office of Science Technology and Policy[6] have documented the challenges faced by dramatic increases in diversity, the challenges of global competition,[7,8] and the criticality of a college education in terms of income and employment.[9]

---

[1] Nick Anderson, "How Supreme Court's Michigan affirmative-action ruling affects colleges," Washington Post, April 23, 2014, http://www.washingtonpost.com/pb/nick-anderson.

[2] U.S. Census Bureau. "U.S. Census Bureau Projections Show a Slower Growing, Older, More Diverse Nation a Half Century from Now," December 12, 2012, http://www.census.gov/newsroom/releases/archives/2012.html.

[3] Michael M. Crow, Differentiating colleges and universities: Institutional innovation at Arizona, *Change: The Magazine of Higher Learning*, September-October 2010.

[4] National Academy of Sciences, National Academy of Engineering, Institute of Medicine (NAS, NAE, IOM), *Rising Above the Gathering Storm: Energizing and Employing America for a Brighter Economic Future*, The National Academies Press, Washington, D.C., 2007.

[5] NAS, NAE, IOM, *Expanding Underrepresented Minority Participation: America's Science and Technology Talent at the Crossroads,* The National Academies Press, Washington, D.C., 2010.

[6] Executive Office of the President, President's Council of Advisors on Science and Technology (PCAST), *Report to the President: Engage to Excel: Producing One Million Additional College Graduates with Degrees in Science, Technology, Engineering, and Mathematics*, PCAST, Washington, D.C., 2012, http://www.whitehouse.gov/sites/default/files/microsites/ostp/pcast-executive-report-final_2-13-12.pdf.

[7] Thomas L. Friedman, *The World Is Flat: A Brief History of the Twenty-First Century*, Farrar, Straus and Giroux, New York, N.Y., 2005.

[8] Public Law 110-69, America Competes Act, August 9, 2007.

[9] Bureau of Labor Statistics, "Employment Projections," http://www.bls.gov/emp/ep_chart_001.htm, accessed May 25, 2014.

In September of 2009, at a briefing session for the House education committee's higher-education subcommittee, David M. Bressoud, president of the Mathematical Association of America, noted that "the 1990s saw significant decreases in both the number of engineering majors and mathematics majors. Both numbers have since recovered, but to only just above the level of 1990 [with] . . . recent recoveries . . . powered almost entirely by white males and non-U.S. residents. Women as well as African, Hispanic, and Native Americans are decreasing as a share of these majors . . . . The number of African Americans earning bachelor's degrees in mathematics was higher in 1992 than in 2007."[10] The lack of diversity among STEM faculty is captured in the observation by Shirley Malcolm, of the American Association for the Advancement of Science, that "mathematics faculties are looking less and less like the student bodies that they are teaching."[11] The Massachusetts Institute of Technology's recent report of the initiative on faculty race and diversity[12] also serves to highlight Shirley Malcolm's observation on faculty diversity. The MIT report and the NRC report on expanding minority participation in STEM education[13] highlight the impossibility of sustained change in faculty diversity without the continuous stream of highly determined young men and women completing baccalaureate STEM degrees at HBCU/MIs.

That NRC report notes that the proportion of underrepresented minorities in science and engineering would need to triple to match their share of the overall U.S. population, revealing a scale of effort that is substantial and that, so far, has been carried out overwhelmingly by HBCUs/MIs. The NRC report notes that the top 25 baccalaureate-of-origin institutions for African American doctorates in the natural sciences and engineering awarded 2,232 doctorates in the period 2002 through 2006; and the top 25 baccalaureate-of-origin institutions for Hispanic doctorates in the natural sciences and engineering awarded 2,038 doctorates in the same period.[14] In fact, the top 10 baccalaureate-of-origin institutions for African American doctorates in the natural sciences and engineering are all HBCUs: Florida A&M University, Howard University, Hampton University, North Carolina A&T University, Spelman College, Morehouse College, Southern University and A&M College at Baton Rouge, Xavier University of Louisiana, Tuskegee University, and Morgan State University. Nine HSIs are ranked among the top 20 baccalaureate-of-origin institutions for Hispanic doctorates in the natural sciences and engineering: University of Puerto Rico Mayaguez, University of Puerto Rico Rio Piedras, University of Puerto Rico Humacao, University of Texas El Paso, New Mexico State University System, Florida International University, University of California Irvine, University of New Mexico System, and the University of Texas San Antonio. Most of these HBCU and HSI institutions have received funding from ARL.

During the century and three-quarters that HBCUs have been evolving, they have been operating at a social and economic disadvantage; they still tend to have smaller endowments and receive less in both government support and private donations than other academic institutions do.[15] They include both public and private institutions, and most are identified as small liberal arts, primarily undergraduate institutions enrolling from a few hundred to a few thousand students. About a quarter of them are universities with a research mission; they include undergraduate, graduate, and first-professional students and award graduate and professional degrees.

---

[10] For information about the September 22, 2009, briefing, see David M. Bressoud, "MAA Speaks Out on Capitol Hill," Launchings blog, October 2009, http://www.maa.org/external_archive/columns/launchings/launchings_10_09.html.

[11] H.G. Grundman, Revisiting the question of diversity: Faculties and Ph.D. programs, *Notices of the AMS* 56(9):1115-1118, 2009.

[12] Rafael L. Reif, *Report of the Initiative for Faculty Race and Diversity*, Massachusetts Institute of Technology, Cambridge, Mass., 2010, http://web.mit.edu/provost/raceinitiative/research-a.html.

[13] NAS, NAE, IOM, *Expanding Underrepresented Minority Participation*, 2010.

[14] Ibid., pp. 252, 259.

[15] Dexter Mullins, "Historically black colleges in financial fight for their future," *Al Jazeera America*, October 22, 2013, http://america.aljazeera.com/articles/2013/10/22/historically-blackcollegesfightfortheirfuture.html.

CHARACTERISTICS OF HBCU/MI INSTITUTIONS

The historic mission of HBCUs has been the education of African American students and advancing the African American community of the United States, especially in the southern, Midwestern, and mid-Atlantic states where they are located. Over time they have become more racially and socioeconomically and culturally diverse, enrolling and employing a diverse population of people from the U.S. and abroad.

Up through the enactment of the Civil Rights Act of 1964 and the desegregation of U.S. higher education, HBCUs enrolled the majority of the nation's black college students, and they produced the majority of degrees awarded to blacks in the United States. Most of the nation's black faculty worked in these institutions, where they represented the majority of the faculty on their campuses. HBCU enrollments and their overall size have grown steadily over time; today they collectively enroll nearly 300,000 students of more ethnic and racial variety than in the past. Although they no longer enroll the majority of black college students or employ the majority of the nation's African American faculty, they enroll and employ a larger share than they account for of students and faculty at the nation's colleges and universities overall. Historically, they have produced graduates who have risen to leadership positions in science and other disciplines and fields.[16]

HSIs represented 6 percent of all institutions of higher education in 2003, enrolling about half of all Latino undergraduates.[17] There are now 370 HSIs, an increase of roughly 60 percent from the 242 colleges that met the definition in 2003.[18] Most HSIs were not originally established to serve a particular student population. HSIs are characterized by their enrollment ratios rather than by their institutional mission. However, there are several exceptions. For example, institutions in Puerto Rico were created with a mission to serve the residents of the island, the vast majority of whom are Hispanic. The oldest of these is the University of Puerto Rico, which was created in 1903. In the contiguous United States, Boricua College (1968) in New York and the National Hispanic University (1981) in California are the only institutions created with the explicit mission to serve Latino students.

Historically, U.S. tribal nations (American Indians, Native Alaskans, or Native Americans) have supported the idea of tribal higher education institutions through their self-determination efforts; individual tribes made significant efforts for the past 100 years to establish such educational entities. In 1978, the Tribally Controlled Colleges and Universities Assistance Act was established, and the ongoing quest for full federal funding began.[19] Tribal Colleges and Universities (TCUs) were created in response to the unique higher education needs of American Indians/Alaska Natives (AI/AN) and generally serve geographically isolated reservation populations that have no other means of accessing education beyond the high school level. The American Indian Higher Education Consortium (AIHEC) noted that "TCUs have become increasingly important to educational opportunity for Native American students and are unique institutions that combine personal attention with cultural relevance to encourage American Indians/Alaska Natives—especially those living on reservations—to overcome the barriers they face to higher education."[20]

---

[16] M. Christopher Brown II and Ronyelle Bertrand Ricard, The honorable past and uncertain future of the nation's HBCUs, *National Education Association Higher Education Journal*, Fall 2007, pp. 117-130, http://www.nea.org/assets/img/PubThoughtAndAction/TAA_07_12.pdf.

[17] *Education Encyclopedia*, "Hispanic-Serving Colleges and Universities–The History of HSIs, HSIs and Latino Educational Attainment, Conclusion," http://education.stateuniversity.com/pages/2045/Hispanic-Serving-Colleges-Universities.html, accessed May 25, 2014.

[18] C. Dervarics, Hispanic-serving institutions continue growth with more poised to join ranks, *Diverse: Issues In Higher Education*, February 25, 2014, http://diverseeducation.com/article/60920/.

[19] American Indian Higher Education Consortium, "History of American Indian Higher Education," February 27, 2000, http://aihec.org/about/documents/AIHECHistorical%20Overview.pdf.

[20] American Indian Higher Education Consortium, homepage, http://aihec.org, accessed May 25, 2014.

A report by the National Academy of Engineering (NAE) noted that "TCUs are similar to other two-year educational institutions in the United States in that they provide access for local students who might not otherwise receive a postsecondary education. Most tribal college students are first-generation students, and their average age, 31.5, is well above the average age of traditional college students. About two-thirds (67 percent) are women . . . and in 2002, more than half were single parents . . . . Overall, 41 percent attend on a part-time basis, although this varies from 84 percent on the Chief Dull Knife campus to 15 percent at the three federally chartered colleges." [21]

VALUE OF ARL SUPPORT TO HBCUs/MIs

The committee's discussions with administrators, faculty, and students at selected HBCUs/MIs, detailed in Chapter 4, overwhelmingly highlighted the positive, sustained impact that ARL limited funding for research to HBCUs/MIs has had in providing inspiring opportunities for URMs, supporting the scientific infrastructure, reshaping STEM curricula, recruiting research faculty, and using the institutions' relationships with ARL to leverage incipient yet critically important relationships with industry and foundations.

Among the HBCUs/MIs with whom the committee engaged in discussion, there are institutions that have moved into the class of research universities. These institutions are now living under a sustained tradition of high-level research and high levels of extramural research funding; faculty that include a mix of national and international individuals who clearly understand, value, and support the mission of HBCUs/MIs; research portfolios that systematically blend graduate and undergraduate participation and are capable of easily accommodating a larger number of students, particularly minority students, if funding were available; recruitment models that take advantage of regional cultures and traditions; partnership with national laboratories and elite institutions; and partnership models that take advantage of the history and commitment of the regional political leadership. These institutions have described success in producing successful STEM undergraduates, enriching top graduate schools programs across the nation. Institutions that are not yet the owners of sustained research programs but that possess a growing core of committed STEM researchers benefited immensely from infrastructure ARL grants and collaborative agreements that maintained or increased their STEM capacity. The limited funding from ARL and other sources has been applied by these institutions to expand their STEM research capabilities by supporting researchers and students. Several institutions with whom the committee held discussions have served the URM communities for decades, have created a culture that attracts top students from underrepresented groups, and have recruited enough outstanding faculty to position them just below the tipping point that moves them into the class of entrepreneurial research institutions. These institutions, now owners of significant extramural funding, have the potential to develop synergistic interactions that transforms them, if the nation invests in them, into the category of research universities.

The role of ARL funding is critical at HBCU/MIs because it is tied in to the importance of building dynamic STEM curricula, and attracting URMs to STEM disciplines in ways that encourage them to enter and complete advanced degrees is fundamentally important. A recent study[22] found that "the effects for URM students were even stronger than for the general population of students. URM students in structured undergraduate research programs were 15 percentage points more likely to enroll in STEM graduate/professional programs. Conducting research with faculty predicted a 14-point increase in URM students' probability of enrolling in STEM graduate/professional programs, and URM students who

---

[21] NAE, "Engineering Studies at Tribal Colleges and Universities—Letter Report from the Steering Committee for Engineering Studies at the Tribal Colleges," The National Academies Press, Washington, D.C., 2006.

[22] S. Hurtado, M. Kevin Eagan, Tanya Figueroa, and Bryce E. Hughes, "Reversing Underrepresentation: The Impact of Undergraduate Research Programs on Enrollment in STEM Graduate Programs," presentation at the annual meeting of the American Educational Research Association, Philadelphia, Pa., 2014.

had any research experience as undergraduates enjoyed a 17-point advantage in their probability to enroll in STEM graduate/professional programs compared to their URM peers who did not have an undergraduate research experience." In the mathematical sciences, the impact of research experiences for undergraduates on encouraging them to pursue advanced degrees in the mathematical sciences or related fields has been extraordinary, particularly when these efforts are carried out in environments where these students are not isolated or underrepresented.[23,24]

Supporting access to and collaborative training in STEM at the graduate and undergraduate levels is central to a national policy aimed at diversifying and increasing the STEM workforce, an area of importance to ARL. An example is provided by the Louis Stokes Alliances for Minority Participation (LSAMP) Program, established in 1991 by the National Science Foundation to develop strategies to increase the quality and quantity of minority students who successfully complete baccalaureate STEM degrees and who continue on to graduate studies in these fields. The LSAMP Program began with grants to six multi-institution alliances. Today, 34 alliances with more than 450 participating institutions have produced thousands of STEM bachelor's degrees. Overall, the LSAMP program has played a key role in increasing the yearly STEM baccalaureate graduation rate of URM students; since LSAMP's inception, the alliance members within the program have produced more than 407,000 URM STEM bachelor's degree recipients.[25] Working within student success programs already in place at these institutions, the LSAMP also provides direct financial assistance to many of its participants. Distinguishing it from traditional scholarship programs, LSAMP takes a multidisciplinary approach to student development and retention, creating partnerships among colleges, universities, national research laboratories, business and industry, and other federal agencies in order to accomplish its goals. Hands-on research experiences and mentoring to build student interest in STEM are among the other important ways in which LSAMP helps.[26]

Extending the reach of LSAMP into the first two years of the Ph.D., a program begun in 2003, is called the bridge to the doctorate (BD). Through the BD program students are supported for 2 years, allowing them to engage with graduate students in master's and doctoral degree programs through monthly meetings and other activities that foster a sense of community among participants that contributes to retention. The program has supported over 2,000 BD students.

HBCUs/MIs have generally received relatively little public funding throughout their history, and their students have largely been from socioeconomically disadvantaged backgrounds. Despite these deficits, however, HBCUs/MIs have persevered to fulfill their mission of educating students, and those with research missions have continually sought support to develop scientific and technological research that would propel them into the mainstream of the nation's research universities. Their participation over the past decade in the ARL grants program has been a vital stimulus for the research and development work of a few prominent engineers and scientists at HBCUs/MIs, offering them advancement and providing their institutions the prospect that over time, if the support is sustained and enhanced, they will join the ranks of the nation's prominent research universities that make important scientific and technological discoveries and allow our nation to lead the world militarily and otherwise.

---

[23] C. Castillo-Chavez and C.W. Castillo-Garsow, Increasing minority representation in the mathematical sciences: Good models but no will to scale up their impact, pp. 135-145 in *Doctoral Educations and the Faculty of the Future* (R.G. Ehrenberg and C.V. Kuh, eds.), Cornell University Press, Ithaca, N.Y., 2009.

[24] C. Castillo-Chavez and C.W. Castillo-Garsow, A preliminary theoretical analysis of an REU's community model, *Problems, Resources and Issues in Mathematics Undergraduate Studies* 23(9):860-880, 2013.

[25] Jamaal Abdul-Alim, LSAMP Program has key role in minority STEM degree attainment, *Diverse: Issues In Higher Education*, August 16, 2012, http://diverseeducation.com/article/17305/.

[26] B.C. Clewell, C.C. de Cohen, L. Tsui, L. Forcier, E. Gao, N. Yung, N. Deterding, and C. West, *Final Report on the Evaluation of the National Science Foundation Louis Stokes Alliances for Minority Participation Program*, The Urban Institute, Washington, D.C., 2006.

4

Generalized Findings from Discussions with Selected Historically Black Colleges and Universities and Minority Institutions

The committee visited and engaged in discussions with administrators, faculty, and students at the following academic institutions: City University of New York, Hampton University, Howard University, Morgan State University, Navajo Technical University, North Carolina A&T State University, Prairie View A&M University, and the University of New Mexico at Albuquerque. A teleconference discussion was conducted with administrators and faculty at the University of Puerto Rico at Mayaguez. Each discussion started off with the affirmation that reported observations would not be attributed to individuals or institutions. The selected institutions had received ARL funding over the last decade. Eight of the institutions are universities with documented involvement in graduate research and the granting of formal STEM degrees for several decades. This chapter of the report briefly summarizes observations derived from the discussions; a more detailed summary is presented in Appendix B.

The selected institutions differ not only with respect to local circumstances but also to the experiences of the faculty involved, which reflect the nature of the research topics and the length and dynamics of their interactions with ARL. Nevertheless, common themes emerged. ARL funding at each institution amounted to at most 10 percent and generally much less of the institution's research budget, although the funding garnered by the individual researchers who participated in the discussions often accounted for a larger percentage.

GENERAL OBSERVATIONS

ARL support of research has had and continues to have a positive, beneficial impact on academic programs, infrastructure, students, and faculty. The basic and applied research described by the institutions was uniformly of high quality and appeared relevant to the Army, DoD, and federal agencies involved in such research. The students who presented their work demonstrated confidence, knowledge, and passion for their work. The supported faculty members were aware of the research issues and advances in their respective fields. Some were highly conversant with the ARL's process and well informed about the ARL organization, its culture, and needs. Researchers at each institution were familiar with the research being sponsored elsewhere by ARL

The faculty involved had done an excellent job of blending graduate and undergraduate participation in their programs, contributing to the overall STEM objectives of the school. The nature of the research being conducted with ARL funding is applicable to the needs and goals identified by ARL, and, using entrepreneurial savvy, faculty had systematically leveraged initial ARL funding to successfully secure DoD, the National Science Foundation (NSF), the National Aeronautics and Space Administration (NASA), or other forms of extramural support to enhance their research capacity and contribute to the institutions' efforts to meet their STEM objectives at the undergraduate and graduate levels.

ARL-funded research has helped to improve community morale and has supported and continues to support expansion of facilities, infrastructure, curricula, internships, research production, and STEM degree production, to an extent consistent with (and often exceeding) the relatively small percentage of the institutions' overall research funding it contributes.

INDIVIDUAL GRANTS AND COLLABORATIVE PROGRAMS

The array of grants and contract vehicle and models used by ARL and ARO over the past decade includes the HBCU/MI funding portfolio. The specific award criteria and management goals vary widely from vehicle to vehicle, described in Chapter 5 and Appendix B, but for the purposes of the current discussion about program type, funding was categorized as individual or group. It was generally observed that funding to a single individual faculty member leads to localized progress and development; group funding, either as stand-alone at one institution or collaborative with other institutions, is more readily incorporated into the local plans for institution building, often manifested in the achievements of centers of excellence. The following observations on individual and group funding were gleaned from the discussions with the representatives of selected HBCUs/MIs.

Discussions at the institutions visited yielded the finding that ARL funding at any given school corresponded to 10 percent or less of its research budget, although it was quite often a greater component of the total funding garnered by the individual researchers who took part in the site visits. All sources of money can have immediate positive impacts: assisting students, enabling research by faculty and students, stimulating new and enhanced course offerings, and setting the stage for further funding by ARL and/or other agencies. On the other hand, individual, short-term grants for research and/or equipment tend to have minimal or no long-term impact on institutional development. *Discussions with representatives of selected institutions and information gathered over multiple committee meetings confirm the value of ARL investment in individual principal investigator grants at HBCUs/MIs, with full understanding of their limited institutional impact beyond the specific research carried out.*

Faculty and administrators emphasized the benefits of securing multi-investigator grants, including collaborative projects funded in connection with the work of colleagues at other research institutions. They had clearly observed the by-products of collaborative dynamics, including the role of collaborative programs in adding fuel that further propels institutional capacity building. The HBCUs/MIs were open to receiving advice and mentoring and collaborating with other institutions, including other HBCUs/MIs. Such larger multiperson, multi-institution programs can be more effective in building the institution, but they must be carefully managed to avoid the perception, expressed during the discussions, that HBCU/MIs are sometimes regarded as second-tier participants by both other research institutions (prime contractors) and ARL staff. It is necessary to treat HBCUs/MIs from the start as full partners in any collaborative enterprise.

Multiperson/multi-institutional grants of extended duration are more likely to have lasting impact on building institutional capability. ARL does offer such support to HBCUs/MIs, and many opportunities for improvement in this area were identified during discussions and are summarized briefly below and in more detail in Appendix A.

THE IMPORTANCE OF SUSTAINED SUPPORT

Faculty and administrators at all the institutions engaged in the discussions were in accord in noting that longer, sustained involvement with an ARL project has advantages over the shorter involvements typical of many single-investigator grants. Although a 3-year single-investigator grant can be an important part of a large, ongoing program, it is not enough by itself to develop a program. If not renewed, the resulting discontinuity in student support may be almost the same as beginning once again. The general sense was that opportunities for contract and grant renewal need to be integrated into the ARL program. *In particular, single-investigator grants to early-career faculty need to be of sufficient duration and, where appropriate, need to include equipment, so that the researchers can address the disadvantage that most HBCUs have of no start-up packages for early-career faculty or, at most, limited such packages.*

FACULTY AND STEM PROGRAM DEVELOPMENT, INCLUDING STUDENT DEVELOPMENT

There was uniformly enthusiastic support for the thesis that funds from ARL had contributed significantly to the development of local STEM programs. ARL funding has been used to carry out research that eventually led to new research centers or important research discoveries or the strengthening of specific groups. These efforts have led to the establishment of new courses, applied Ph.D. programs, and new curricula at the undergraduate and graduate levels.

In many cases, the faculty have been innovative in blending graduate and undergraduate participation in their programs, using various means, including an opportunity to earn university credits for participating in research. The underlying idea is to start the pipeline for research early. The faculty believe that more minority students would move on to the Ph.D. program rather than seek immediate employment (e.g., in industry) if they were exposed to research as part of the undergraduate program.

The faculty generally believe that ARL funding has supported twofold benefits in their STEM programs. First, it fosters the interaction that transpires between the graduate students and the undergraduates. An informal communication network often develops in which students engage in dialogue and contribute time to help one another and to elicit insight into the nature of the research being conducted. Second, the funds permit the faculty to engage in dialogue with ARL counterparts as well as those at other universities. However, it was also noted that the ARL dialogue is not frequent or robust enough and that improvement is needed in this area by the ARL leadership. It was frequently observed that ARL needs to work with local faculty to develop an effective plan and process that increases the technical interaction between ARL researchers and university faculty and students.

EQUIPMENT AND FACILITIES

One may generalize about the lack of competitive facilities at HBCU/MIs relative to larger research universities. Several discussants indicated that equipment shortfalls of individuals (particularly those in start-up positions) and of multiusers in facilities were greater than those of top-ranked competitors/collaborators and that the small relative size of HBCU-related grants made equipment purchase and development difficult. In addition they noted that in many instances, minimal funds were available from university sources to provide maintenance and upgrades on equipment that had been painfully acquired. Important exceptions to this situation could be found in those centers of excellence that had grown over the years and were often supported by multiyear, collaborative grants.

This lack of extensive infrastructure gives rise to a disadvantage in the competition for more equipment, a Catch-22 situation. For example, equipment proposals from an HBCU are typically looking for fairly rudimentary equipment (new hires do not normally receive a start-up package; state funding yields buildings, but rarely equipment), while the proposal from a major university will evince existing infrastructure. Lack of technical assistance and minimal matching funds may also have an impact here. Furthermore, even if a proposal is successful, lack of continuing funding or related research funding may make a piece of equipment at an HBCU less effective after several years than it might have been at a large institution. All of these factors may lead ARL program managers to expect a lower potential return on an investment in an HBCU.

ADMINISTRATION OF AWARDS

The process of initiating proposals and managing grant and/or contract funding can vary markedly from one institution to another. Universities farther along on the research university development path will have strong central systems to support individual or group faculty efforts. By contrast, many HBCUs still find themselves in the situation where individuals must go it alone, without a strong institutional support system, as they deal with the highly varied elements and processes of award administration, which differ not only between grants and contracts at ARL, but also from one funding agency to another. Administrative issues vary widely from one school to the next. Several specific examples are explored in Appendix A.

STUDENT ENGAGEMENT WITH ARL

An ARL goal for the HBCU/MI program is that students will become aware of and, to the extent possible, participate in Army programs. The funding and quality of research provide a relevancy that the students appreciate. While one unstated goal of direct student involvement is eventual employment at ARL to help address issues of workforce diversity, this outcome is neither anticipated nor measured by ARL. Faculty and students interviewed were well aware of the source of funding and, in most cases, the relevance of the research to Army goals (or at least to a stated ARL research agenda). However the student experiences varied widely from school to school and from one type of funding entity to another.

One important factor is the country of origin of the graduate student. At some of the schools visited, foreign students are in the majority, while in others they are not. Access to the ARL campus may be quite difficult to achieve for those foreign national students who are not U.S. citizens, limiting their participation.

Another significant issue is geography. It is not surprising to find that student visitation and direct participation in ARL activities are more readily supported for those close to the ARL facilities, especially in view of tight travel budgets.

Some of the responses from the institutions revealed the highly diverse character of student interaction opportunities with ARL. Summer internships at ARL present one very specific opportunity for direct student–ARL interaction, but they were viewed from two opposing perspectives by faculty. In one case an ARL intern characterized the internship as a great experience, especially for those who want to keep research connections with ARL, perhaps as a permanent hire. From another perspective, a summer internship, though valuable, could disrupt a student's progress toward completing a project task, thesis, or dissertation, adding to the time it takes to earn the degree. Discussion considered the value of connecting the thesis/dissertation research to internship tasks at the beginning of a funded project so that time would not be lost. This would take some attentive planning on the part of the ARL program manager and the funded institution.

ARL–UNIVERSITY INTERACTIONS

Among the ARL goals for its HBCU/MI program is increased interaction between ARL staff and the faculty and students at the universities. This interaction could take many forms. The closest might be cooperation in establishing program goals, followed by direct collaboration at the ARL facilities and/or by the exchange of data, specimens to characterize, and other research information or samples. An intermediate level of collaboration might involve regular communication during performance of the grant/contract, with less involvement in project planning and less coordinated sharing of data, information, and samples. Minimal interaction might be limited to infrequent communication, primarily aimed at making the university personnel aware of the intended Army goals in sponsoring the stated research.

Staff at all interviewed institutions expressed a desire for ARL interaction; administrators and faculty frequently mentioned their desire for more frequent, direct contact with ARL and voiced some frustration that this was not happening.

The dynamics for early-career faculty often involve senior faculty who have the right contact(s) at ARL/ARO. Well-connected senior faculty play the role of finders of opportunities and mentors before and during the merging of interests. Many suggested that this process might be significantly modified if serious local capability addressing Army needs and goals is the desired result at the HBCUs/MIs.

Many faculty members expressed the desire for greater participation in the development of the BAAs prior to their issuance. This would imply a very different relationship with ARL—namely, one in which strategic goals were mutually developed and shared. It would also be beneficial if ARL were to provide more detailed and timely reviews of research white papers submitted by HBCU/MI researchers. Specific comments on the quality of proposed research would be welcomed by local researchers. More feedback from ARL program managers on failed white papers and proposals would help faculty target for success.

ARL mentoring could also be applied to help find other elements of the Army that could fund the continuation of research work related to the original ARL funding—for example, the Army medical laboratories or other organizations within the Army Research, Development and Engineering Command, the home of ARL. It appeared also that single PIs could receive useful mentorship on connecting with other, non-ARO divisions of ARL for follow-on research support. A related discussion focused on assisting HBCU/MIs to get access to and time on Army/DoD equipment and facilities, including the Army's major shared resources, such as supercomputing centers.

5

Army Research Laboratory Programs for Historically Black Colleges and Universities and Minority Institutions

In its description on the Internet of its partnership programs, the Army Research Laboratory (ARL) notes as follows:

> With the current pace of technology advancement, insular research and development (R&D) organizations will rapidly lose relevance and value. ARL has adopted business practices that have created a collaborative research environment between it and the private sector in select technology areas. ARL has also provided the Army access to private sector sources of research with the requisite diversity and quality. Currently, ARL outsources 80 percent of its research program to academia with over 250 academic partners in all 50 states and to industry through a mix of grants, cooperative agreements, other transactions authority, or contracts.[1]

In addition, ARL has continued to support single principal investigator (PI) projects, a tested model for the identification of leading researchers with entrepreneurial talents and skills.

ARL SINGLE PRINCIPAL INVESTIGATOR PROGRAMS

Single ARL principal investigators from HBCUs/MIs can significantly enhance the research mission of their institutions. Successful single-PI efforts—those that result in peer judgment of high-quality research—facilitate the identification of scientific leadership, scientific entrepreneurial talent, and inspirational skills that are critically needed at all institutions of higher learning. Single-PI projects facilitate, enhance, promote, and support interactions between students and researchers; this is a very successful model of mentorship in STEM. ARL has used roughly two-thirds of its HBCU/MI funding to support single-PI efforts.

A recent article[2] observed that

> College graduates had double the odds of being engaged at work and three times the odds of thriving in Gallup's five elements of well-being if they had had 'emotional support'—professors who 'made [them] excited about learning,' 'cared about [them] as a person,' or 'encouraged [their] hopes and dreams.' Graduates who had done a long-term project that took a semester or more, who had held an internship, or who were extremely involved in extracurricular activities or organizations had twice the odds of being engaged at work and an edge in thriving in well-being.

---

[1] Army Research Laboratory, "HBCU/MI," http://www.arl.army.mil/www/pages/9, accessed May 25, 2014.

[2] Scott Carlson, "A Caring Professor May Be Key in How a Graduate Thrives," *Chronicle of Higher Education*, May 6, 2014, http://chronicle.com/article/A-Caring-Professor-May-Be Key/146409/?cid=at&utm source=at&utmmedium=en.

The impact that HBCUs/MIs are having in training the next generation of STEM Ph.D.'s relies on the fact that students attending these institutions have access to cutting-edge research. ARL single-PI funding not only brings research of interest to the Army into HBCUs and MIs but also, by putting funds in the hands of single PIs, reinforces the value, importance, and impact of individualized mentorship models. Single-PI ARL funding has increased the capacity for providing research opportunities at HBCU/MI institutions, naturally benefitting URMs.

The ARL resources provided to successful researchers—those who achieve peer judgment of high-quality research—have a direct impact on PI research programs, contributing to the development of the scientific workforce by offering training opportunities from the undergraduate to the postdoctoral level. Successful single PIs are entrepreneurs, aggressively searching for resources to carry out research programs that create new knowledge, enhancing the research capacity of the institution. Successful PIs were identified at all the institutions with which the committee engaged in discussion.

Navigating through the ARL funding process does not appear to be straightforward. PIs with prior knowledge of the culture of ARL and familiarity with the outreach efforts of specific ARL program directors became ideal mentors of early-career researchers looking for ARL funding. The lack of synchrony between the time to a Ph.D. and the duration of the funding (3 years) can limit the ability of HBCUs/MIs to support Ph.D. candidates.

ARL COLLABORATIVE PROGRAMS

ARL is engaged in a variety of agreements with HBCUs and MIs that provide opportunities to participate in Defense research programs. The ARL funds STEM research programs at the institutions, STEM programs for students, and programs and opportunities for faculty and staff to interact with ARL scientists and engineers, to access scientific and technical information, and to collaborate with other educational institutions or research facilities such as the DoD laboratories. ARL programs are highlighted in Figure 5.1 and described in Appendix B. Of note, only two programs focus specifically on HBCU/MI support: HBCU/MI ARO Core Grants and the Partnership in Research Transition (PIRT), though support may also be provided through the other ARL programs.

RECONSIDERING THE BALANCE BETWEEN COLLABORATIVE PROGRAMS AND SINGLE-PI PROGRAMS

The ratio of number of grants to number of collaborative programs is high, averaging 87 percent to 13 percent over the past decade (see Table 5.1); in terms of real dollars, the ratio is lower—approximately 2 to 1 (see Table 5.2). Collaborative programs have received nearly 33 percent of the funds allocated to HBCUs/MIs over the past 10 years.

Collaborative Programs Foster Institution Building

As noted in Chapter 4, the discussions with representatives of HBCUs/MIs confirmed that all sources of money, whether grants or cooperative agreements, have immediate positive impacts, assisting students, enabling research by faculty and students, stimulating new and enhanced course offerings, and setting the stage for further funding by ARL and/or other agencies. *Many of the administrators and faculty members interviewed suggested that cooperative/collaborative programs with ARL are more effective than single-PI grants or contracts in achieving the objective of institution building, provided that the cooperative/collaborative programs are properly managed.*

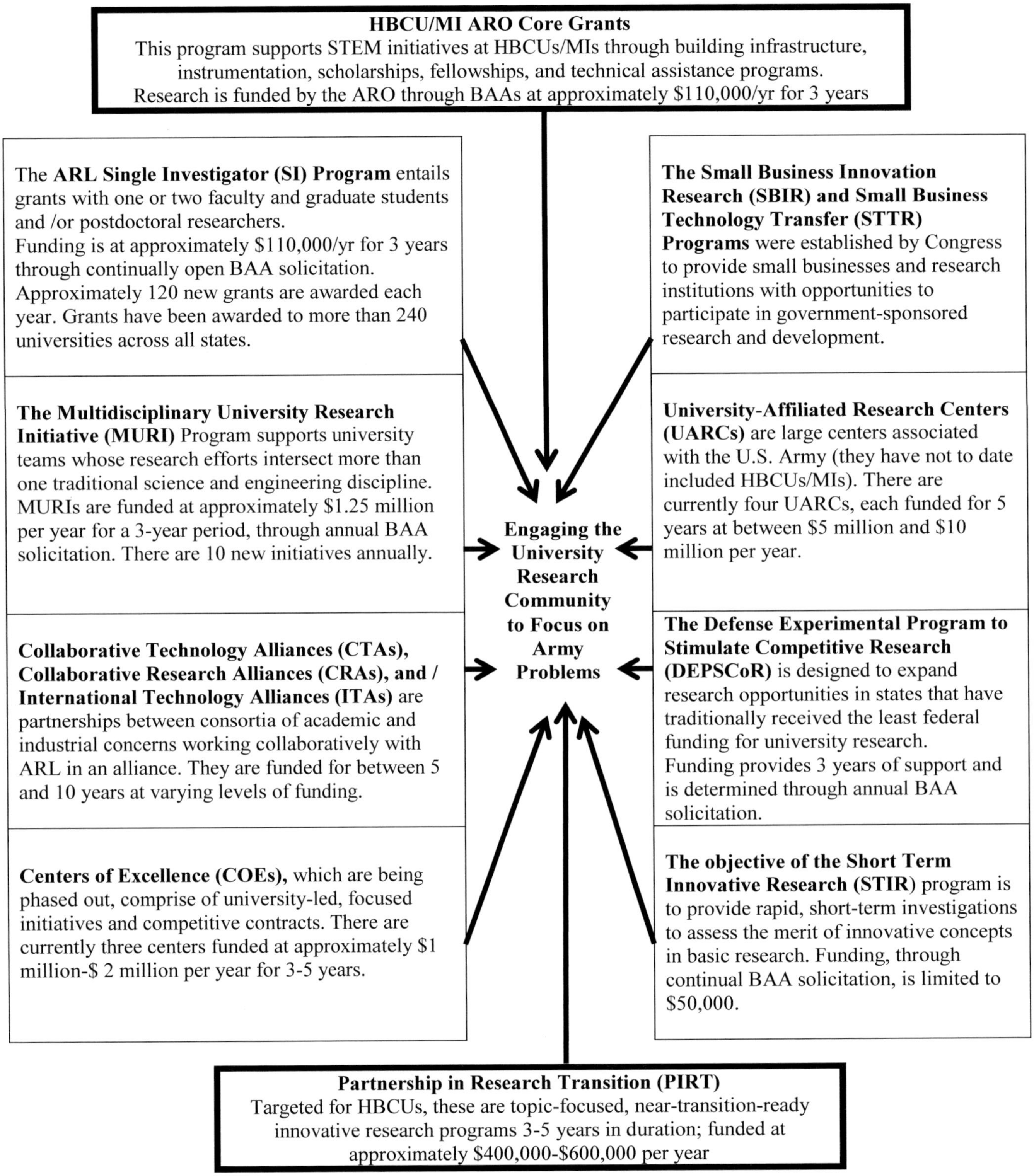

FIGURE 5.1 ARL programs that engage the university research community to focus on Army problems. (Based on a chart presented to the committee by the manager of the ARL Outreach Office.) HBCU/MI ARO core grants and PIRT programs specifically target HBCUs/MIs, although other programs may also involve HBCUs/MIs.

TABLE 5.1 Number and Share of Grants and of Cooperative Agreements

Year	Number of Grants	Number of Cooperative Agreements	Percentage of Grants	Percentage of Cooperative Agreements
2004	98	13	88	12
2005	114	17	88	12
2006	128	20	84	16
2007	126	18	86	14
2008	121	17	86	14
2009	83	9	89	11
2010	106	3	97	3
2011	90	9	90	10
2012	116	17	85	15
2013	90	18	80	20
Total/Average	1,093	141	87	13

TABLE 5.2 Comparison of Funds for Grants and Funds for Cooperative Agreements

Year	Grants (%)	Cooperative Agreements (%)
2004	76	24
2005	76	24
2006	58	42
2007	53	47
2008	51	49
2009	69	31
2010	84	16
2011	61	39
2012	67	33
2013	70	30
Average	67	33

The interviewees suggested that cooperative/collaborative programs are more effective because they

- Permit more extensive person-to-person interactions, which are more valuable in the long term for the students, faculty, and institutions.
- Facilitate the mentoring of fledgling institutions by more experienced HBCUs/MIs. A novice institution would benefit from securing collaborative projects funding with colleagues in

other more established research institutions that could add fuel to propel the capacity building of the institution.

- Enhance continuity. A 3-year grant can be an important part of a large, ongoing program, but a single-PI grant is not enough by itself to develop a program. If a grant is not renewed, the resulting disruption may require the investigator to start a new funding process.
- May help institutions in building their desired infrastructure and institutional base.
- Enhance the stability of an institution's research programs, reduce the uncertainty of graduate student participation, and provide better research refocus at the institutions.
- Represent a more strategic approach to identify and reinforce the already growing centers of excellence at some HBCUs, building on their strengths.

The following factors were suggested by the interviewees with respect to the need for proactive management of cooperative/collaborative programs:

- When HBCU/MI involvement in programs is mandated, or when their participation is invited as an afterthought by the prime contractors, the HBCUs/MIs may not be given adequate consideration during program planning and development. For example, a prime contractor may organize a proposal and then identify the HBCU that "fits" the already written proposal. In such a case, little may be expected from the HBCU/MI participants, and little effort may be expended by either the prime or ARL program managers to draw upon the full capabilities of the HBCU/MI participants.
- In multiperformer projects, participating HBCUs/MIs often receive a much smaller portion of the project tasking than their non-HBCU/MI collaborators and much less research funding and support dollars. In some cases, lead non-HBCU/MI institutions may bring HBCU/MI students to the lead institution as research participants; this will benefit these students, but it will have no lasting impact on the HBCU/MI faculty, infrastructure, or reputation.
- HBCUs/MIs sometimes feel that they are not participants in the critical decision-making paths of consortia and that their science and technology (S&T) capabilities are not adequately utilized by the prime contractors.
- As regards UARC participation, HBCU/MI funding is not substantial enough to create the strong and sustaining foundation needed to compete successfully for UARC projects year after year without ARL mentoring and interaction.
- Without diligent management by ARL, program leaders may reallocate funding originally slated for HBCU/MI participants.

HBCU/MI faculty and administrators interviewed were consistent in suggesting that spreading limited funds across a large number of individual PIs and numerous HBCU/MI institutions provides positive short-term benefits to faculty and the students but fails to consider the long-term development of institutional capability.

Acknowledging that most HBCUs/MIs would not be where they are without grant programs like those funded by ARL, the HBCUs/MIs may be at an inflection point where revisiting the ratio of single-PI grants to collaborative programs is warranted, with an eye toward institution building. Increasing the allocation of funds to cooperative/collaborative programs with significant HBCU/MI participation could encourage the institutions to generate better strategic plans for growing their capabilities and for serving their student population.

Considering the Impact of Funding Approaches

ARL needs a strategic plan for the allocation of funds to HBCUs/MIs that includes assessment of the impact of the funds in terms of HBCU/MI program goals set by ARL (including, possibly, the goal of institution building) at those institutions and that applies the assessment to appropriately balance single-PI grants and collaborative/cooperative programs. Depending on how ARL decides to allocate funding to experienced, highly successful HBCUs/MIs, helping them to progress to the point at which they do not depend on such support, and to fledgling HBCUs/MIs, helping them to begin institution building, rebalancing will have to be slowly orchestrated.

Historically, as indicated in Table 5.3, over the past 10 years approximately 74.2 percent of the dollars and 70.9 percent of the cooperative agreements went to four institutions. The implication is that some HBCUs/MIs could mentor others on how to develop a productive, institution-building relationship with ARL. ARL could also learn from the successful HBCUs/MIs the success strategies that they have developed.

TABLE 5.3 Cooperative Agreements at 14 HBCUs/MIs

HBCU/MI	Share of Total HBCU/MI Funding (%)	Share by Number of Agreements (%)
New Mexico State University	46.8	28.37
North Carolina A&T State University	15.0	24.11
Howard University	8.4	11.35
Tennessee State University	4.0	7.09
Subtotal	74.2	70.90
Tuskegee University	4.8	6.38
Hampton University	4.3	4.96
Prairie View A&M Research Foundation	3.8	4.96
Florida A&M University	5.7	2.84
Morgan State University	0.4	2.84
Clark Atlanta University	0.5	1.42
Lincoln University at Jefferson City	2.4	1.42
City College of New York, Queens College	0.1	1.42
University of Puerto Rico at Mayaguez	0.1	1.42
Total	100.0	100.00

TABLE 5.4 HBCU/MI Shares of Total ARL Funding for Single PIs (percent)

	Dollar Amount of Grants	Number of Grants
University of California, Riverside	6.6	10.0
University of New Mexico, Albuquerque	9.8	9.7
University of California, Berkeley	23.5	8.8
University of California, Los Angeles	5.9	8.2
North Carolina A&T State University	6.5	6.2
University of Puerto Rico at Mayaguez	7.2	6.2
City College of New York	2.9	5.0
Howard University	2.2	4.1
University of California, Irvine	2.5	3.8
Florida A&M University	1.9	3.3
City College of New York (Flushing)	1.6	2.9
University of Texas at San Antonio	1.7	2.6
Total	72.3	70.8

As indicated in Table 5.4, several institutions are very experienced in securing single-PI grants. These institutions appear to have learned well the proposal development process and have overcome administrative roadblocks, so they might be good mentors for fledgling institutions.

It is necessary for ARL to be proactive in securing the significant participation of HBCUs/MIs in multiyear cooperative agreements to ensure that these institutions have adequate funding and time to gain access to and procure equipment, support the completion of graduate student research, arrange for onsite or virtual internships with ARL laboratories or other laboratories, and develop the capacity to respond to ARL programmatic redirection of funded research tasks. As part of a proactive strategy, it is important that ARL consider the following:

- Communicating a strategic vision for the program that aims to enhance the science and engineering education beyond just funding (e.g., internships, symposia),
- Assisting in mentoring the HBCU/MI candidates responding to the BAA and executing tasks after award,
- Reviewing its core and cooperative agreement BAA processes to minimize administrative burdens on HBCU/MI respondents.
- Developing metrics of performance to measure contract performance,
- Developing metrics for the impacts of the funding on the institution building of the recipients,
- Ensuring meaningful participation of fledgling HBCUs/MIs, and
- Assisting successful institutions to progress to a level of institution building that allows them to compete on an even footing with other research institutions.

6

Conclusions and Recommendations

Overall, the Army Research Laboratory (ARL) programs supporting historically black colleges and universities and minority institutions (HBCUs/MIs) are strong, well run, and commendable. These programs have, over many years, provided support to many HBCU/MI individual researchers and institutions, administered through a variety of programs accessible to HBCU/MI researchers and institutions and reported by recipients to be helpful in their development and expansion of STEM programs. Within the narrow confines of the committee's tasking—that is, examining institutional science, technology, engineering, and mathematics (STEM) improvements at HBCUs/MIs as a result of ARL funding—the strong program can be made even stronger. The recommendations that follow are presented in that spirit.

KEY CONCLUSIONS AND RECOMMENDATIONS

Neither the Army nor, specifically, ARL has put in place written directives or a strategic plan for supporting HBCUs/MIs or for assessing the impacts of that support.

1. ARL should create and disseminate a policy directive regarding its commitment and priorities for a credible and sustainable HBCU/MI support program. Specifically, ARL should articulate a vision and write a strategy to enhance STEM capability within its HBCU/MI-supported community and develop metrics to measure STEM capability improvement; metrics should include progress toward independence, including expanded funding relationships with other funding agencies.

It is important that HBCU/MI participants are fully and successfully engaged in programs within which they participate. On that account, it is important that ARL solicit and receive regular feedback from participants with respect to their experiences. When HBCUs/MIs are subcontractors to non-HBCU/MI institutions on ARL programs, the prime contractors are often relied on to assign and monitor tasks performed by the HBCUs/MIs. It is important that ARL maintain cognizance and control over the task assignment and monitoring tasks to assure that funding is equitable and that the tasking capitalizes on HBCU/MI capabilities. Onsite collaboration of ARL researchers at HBCUs/MIs would help to facilitate ARL mentoring opportunities and to encourage communication between HBCU/MI researchers and ARL

2. ARL should examine the funding of collaborative projects involving HBCUs/MIs and non-HBCUs/MIs to ensure that the funding is equitable and that the tasking takes advantage of HBCU/MI capabilities. ARL should require HBCU/MI participants in ARL-funded collaborative, cooperative agreement projects to provide to ARL regular reports on their experiences with the project planning, execution, management, funding, and other collaborative interactions with the sponsoring ARL program manager and other participants in the collaboration team.

Once ARL has established its strategy to enhance STEM capability at HBCUs/MIs and metrics to measure STEM capability improvement, it can examine the impacts of the institutions' use of ARL support funds and adjust funding to encourage desired impacts.

3. ARL should regularly assess which HBCU/MI activities have the most successful impact on the development, maintenance, and growth of their STEM programs and should rebalance funding according to those assessments.

Over the past 20 years, many minority faculty have left HBCUs/MIs and joined nonminority institutions, and now many recipients of ARL HBCU/MI funding are not African American, Hispanic, or Native American. Underrepresented minority (URM) funding programs exist at other federal agencies, including the National Institutes of Health (NIH) and the National Science Foundation (NSF), as well as at the Department of Defense (DoD).

4. ARL should consider NIH, NSF, and other URM funding incentive models in allocating support, from within its HBCU/MI funds, for URM U.S. citizen undergraduate and graduate research students, summer interns, postdoctoral fellows, and faculty researchers.

It is important that supportive funding for HBCUs/MIs be sustainable. The overwhelming proportion of ARL funding for HBCUs/MIs has been through grants to individual investigators. While there is value in this process, there is also great value in participation by HBCUs/MIs in multiyear cooperative agreements, which offer broader and longer-term support for building institutional STEM capabilities. HBCUs/MIs would benefit from ARL support to secure and successfully participate in multiyear cooperative programs.

5. ARL should proactively engineer the participation of its HBCUs/MIs in multiyear cooperative agreements to ensure that there is adequate funding and time for those institutions to gain access to and procure equipment, support the completion of graduate and undergraduate student research, arrange for onsite or virtual internships with ARL laboratories and other laboratories, and develop the capacity to respond to redirection of funded research tasks by ARL program managers. As long as ARL continues its University-Affiliated Research Center (UARC) programs, it should regularly consider HBCUs/MIs for UARC designation or for formal partnerships with existing Army UARCs based on continual adequate technical performance and extant STEM talent and physical capabilities.

The process of initiating proposals and managing grant and/or contract funding can vary markedly from one institution to another. Those universities farther along on the research university development path will have strong central support systems to support individual or group faculty efforts. By contrast, many HBCUs/MIs still find themselves in the situation that individuals must go it alone in dealing with the highly varied elements of award administration.

6. With HBCUs/MIs in mind, ARL should review its core and cooperative agreement Broad Agency Announcement (BAA) processes to minimize administrative burdens on university respondents and should, with input from HBCUs/MIs, consider mentoring opportunities to enable more awareness of and success in responding to BAA opportunities.

As noted above, the overwhelming proportion of ARL funding for HBCUs/MIs has been through grants to individual investigators, which, without deliberate efforts to connect these individuals to multiinstitutional research teams, can result in loss of synergistic opportunities that such collaborations encourage.

7. ARL should gather best practices from other agencies in order to design models of funding that systematically connect and integrate single principal investigator research at HBCUs/MIs, with the efforts carried out by multi-institutional research teams, to facilitate institutional building and the development of entrepreneurial scientific leadership at the HBCUs/MIs.

HBCUs/MIs themselves need to encourage in their researchers and administrators wider recognition of the opportunities available for collaborative interactions and the value of such interactions for institutional STEM building, and they need to seek out more of those opportunities.

8. The HBCUs/MIs should pursue more ARL-supported collaborative research funding. Led by faculty and institutional leadership, HBCUs/MIs should engage in research opportunities that include collaborative grants and contracts as well as single investigator research and development.

To be consistently successful, researchers require the support of institutional administrative offices that perform and assist with such tasks as identifying sponsored opportunities, serving as liaisons with sponsoring agencies and collaborating institutions, following procedures for preparation of proposals, administering financial and other contractual aspects of programs, and supporting student assistants and researchers. These tasks are performed best by professional administrators and can become a counterproductive burden that may not be performed efficiently and effectively by the researchers themselves.

9. The HBCUs/MIs should continuously improve the efficiency and effectiveness of their offices of sponsored programs that assist their faculty in execution of ARL-supported research programs both on and off campus.

ARL is not the sole source of potential funding for HBCUs/MIs, of course. ARL support can be a valuable element within a wider strategy of STEM building that includes partnerships with or support from other sources, including industry and foundations based both here and abroad as well as federal agencies.

10. The HBCUs/MIs should expand their ARL-supported research by partnering with local industry and international sources. The HBCUs/MIs should build relationships with foundations wherein foundational resources are combined with ARL resources to extend the research portfolios of the HBCUs/MIs.

ADDITIONAL CONCLUSIONS AND RECOMMENDATIONS

ARL does announce sponsoring opportunities through such traditional channels as the Broad Agency Announcement, with which past recipients of ARL support are familiar. HBCUs/MIs would benefit from additional outreach mechanisms designed to inform potential applicants of available opportunities and of the procedures for becoming involved in ARL-funded programs.

11. ARL should consider sponsoring periodic information dissemination symposia with the dual purpose of featuring the products of HBCU/MI-sponsored research and development and facilitating the active networking and collaboration among the research community.

As noted above, ARL is one of multiple potential sponsors of STEM research at HBCUs/MIs, which have been encouraged to seek and combine resources from additional sponsors to enhance institutional STEM capabilities. By collaborating with other agencies, ARL may assist this strategic effort.

12. ARL should consider collaborating with agencies inside and outside of DoD to seek joint sponsorship of projects as a means of expanding the support for research and development at HBCUs/MIs.

Dissemination of research results and impacts is an important step in the research and development process. Public awareness and appreciation of HBCU/MI achievements under ARL sponsorship provides benefits to the institutions and to ARL, including the attraction of other HBCUs/MIs to the ARL programs, enhancement of the visibility and stature of the performing institutions and researchers, and attraction of researchers and students to the HBCUs/MIs.

13. ARL should seek ways to provide public recognition for the sponsored research of HBCUs/MIs, especially in the state and local communities where the institutions are located.

Providing financial support in the form of tuition, room and board, books, and supplies is one means to support students at HBCUs/MIs who serve as research assistants. Encouraging and helping students to interact with other students and researchers at their institutions, at other institutions, including industrial and government institutions, and at ARL would expand their professional development. ARL can, with attention, identify means of expanding these and perhaps other forms of support.

14. ARL should seek ways to expand the support at HBCUs/MIs of students serving as research assistants on ARL-sponsored projects.

Unless single principal investigator grants are for a sufficient time period, they will not support the activities required for involved graduate students to achieve their degrees.

15. ARL single principal investigator grants and collaborative/cooperative agreements should be multiyear selections (with a 5-year norm) to enable graduate student thesis/dissertation success and to adjust to in-term ARL program direction changes.

The BAA is a primary mechanism by which ARL announces its funding opportunities for HBCUs/MIs. Mentoring by ARL of applicant candidates would facilitate their effective participation in the proposal process, and additional instruction in other funding opportunities would help them to expand their potential funding sources.

16. ARL should proactively and regularly mentor HBCU/MI candidates for funding on the BAA-selection-program execution process and on identification of other funding and collaboration opportunities within the Army.

HBCU/MI interactions with and mentoring by other institutions, including other HBCUs/MIs and other DoD agencies, would facilitate the sharing of information, including examples of successful practices. This would encourage beneficial collaborations that help to enhance STEM institutional development as well as individual growth.

17. ARL should create a more comprehensive, proactive strategy for providing exposure of HBCUs/MIs—including their faculty, students, research, and the overall institution—to other HBCUs/MIs and other DoD agencies, and should promote formal discussion venues that link experienced program participants with HBCUs/MIs.

Appendixes

A

Observations Gleaned from Discussions with Representatives of Selected Historically Black Colleges and Universities and Minority Institutions

The committee visited and engaged in discussions with representatives of the following academic institutions: City University of New York, Hampton University, Howard University, Morgan State University, Navajo Technical University, North Carolina A&T State University, Prairie View A&M University, and the University of New Mexico at Albuquerque. A teleconference was conducted with administrators and faculty at the University of Puerto Rico at Mayaguez.

The face-to-face meetings involved primarily "around the table" information-gathering sessions with members of the universities' research leadership teams and key administrators, faculty that have received funding from ARL or related sources, and undergraduate and graduate students supported by the Army Research Laboratory (ARL) or related sources of funding. Each session began by assuring participants that no observations would be attributed to either individuals or institutions.

The nine institutions selected for visits include, to the extent possible, institutions that had received enough aggregate funding over the last decade to warrant the expectation that the information gathered during these visits would assist the committee in identifying findings and drawing broad observations leading to the recommendations expected from its work. Consequently, eight of the nine discussions took place at universities with documented involvement in graduate research and in the granting of formal science, technology, engineering, and mathematics (STEM) degrees for several decades. These eight institutions shared the general status of "good to great" research institutions in select areas mostly driven by the efforts of individuals or small groups; institutions with average to poor reputations in those research areas were not targeted. This appendix summarizes observations made during discussions with these eight institutions and one institution that is only now making a transition from a 2-year associate degree program to a full 4-year undergraduate program.

The eight institutions differ not only with respect to the local circumstances but also to the experiences of the faculty involved, which—as expected—are reflected in the nature of the research topics and the length and dynamics of their interactions with ARL. Nevertheless, common themes emerged. This appendix first collects strong overall impressions. It then proceeds to focus on selected broad areas of discussion, with an emphasis on current status, and highlights opportunities for improvement. Central to the discussions were the issues of scale and impact of ARL funding in the selected institutions: At any given school such funding made up 10 percent or less of its research budget, although it quite often formed a greater proportion of the total funding garnered by the individual researchers who took part in the site visits.

GENERAL OBSERVATIONS

ARL support of research has had and continues to have a positive, enhancing impact on academic programs, infrastructure, students, and faculty. The research appeared to be relevant to the Army, the Department of Defense (DoD), and the federal agencies involved in support of the basic and applied research generated by challenges as identified by the ARL research leadership. The students presenting their work demonstrated confidence, knowledge, and passion for their work. The ARL-supported faculty members demonstrated an awareness of the research issues and advances in their respective fields. Some were highly conversant with the ARL's process and well informed about the organization, its culture, and its needs. Researchers at each institution were familiar with the research being sponsored elsewhere by ARL.

The research viewed was of high caliber. It was clear that the faculty involved have done an excellent job of blending graduate and undergraduate participation in their programs, contributing to the overall STEM objectives of the school. While the nature of the research being conducted with ARL funding is highly applicable to the needs and goals identified by ARL, the faculty demonstrated entrepreneurial savvy that systematically leveraged initial research ARL funding to successfully secure extramural support from DoD, the National Science Foundation (NSF), or the like. Such support was not only enhancing their research capacity but also contributing to meeting their institution's STEM objectives at the undergraduate and graduate levels.

ARL-funded research has had and continues to have a positive impact on facilities, infrastructure, curricula, internships, research production, STEM degree production, and community morale to an extent not only consistent with but also often exceeding the relatively small percentage of overall research funding ARL provides.

Faculty and students appreciate the funding received from ARL. In the best of circumstances, ARL funding has contributed to the capacity building of the science and engineering education programs by attracting high-quality and productive research scientists to the faculty; by contributing to the production of high-quality and diverse Ph.D. scientists; by contributing to the stature of the university, helping it to fulfill its research university status goals; and by expanding the number and quality of STEM opportunities, research experiences, and degrees available to undergraduates.

In many instances the aggregate of ARL funding over many years has supported and enabled the advancement of curricula, the creation of new academic departments, the improvement of business practices and implementation of models of compliance, the start-up of high-quality laboratory facilities, the establishment of focused trails of research publications, and the systematic support of undergraduate and graduate students in STEM fields.

INDIVIDUAL GRANTS VERSUS COLLABORATIVE PROGRAMS

The array of grants and contract vehicles and models used by ARL and the Army Research Office (ARO) over the past decade includes the historically black colleges and universities/minority institutions (HBCU/MI) funding portfolio. While the specific award criteria and management goals vary widely among the identified vehicles, for the purposes of this report the committee divided the vehicles/models into two categories, funding for individuals and funding for groups, to help in organizing the discussion about program type. It was generally observed that funding a single individual faculty member leads to localized progress and development, while funding a group, either as stand-alone at one institution or in a collaboration with other institutions, is more readily incorporated into local plans for institution building. This observation is often borne out by the success of centers of excellence.

All sources of money can have positive impacts, some of them immediate, others longer term: assisting students, enabling research by faculty and students, stimulating new and enhanced course offerings, and setting the stage for further funding by ARL or other agencies. Individual grants for research or equipment with short-term duration tend to have minimal or no long-term impact on institutional development. Indeed, while ARL's principal investigator grants are valuable, their impact is generally limited to the specific research carried out. Collaborative programs with ARL were considered by the administrators and faculty preferable to such individual grants or contracts because they lead to person-to-person interactions that are more valuable in the long term. The ARL/ARO single principle investigator (PI) program was considered extremely positive. Single-PI funding was seen as complementary, especially for shorter time frames, to projects that would occupy the time of a professor and at most one or two master's students. The amount of funding and the time frame of single-PI grants defined its scope and limitations. Larger collaborative project funding that could extend beyond 3 years and be substantial enough to engage a team is important for professors and Ph.D. candidates; such funding allows them to tackle complex problems. The faculty sees the cooperative programs as providing the following: diversity of technical input, leading to more innovation; opportunity to network with faculty and students within and between institutions; increased student exposure that leads to internships; lower overhead allocation (only 23 percent overhead on funds to subcontractors); and opportunity to enhance a university's reputation and its attractiveness to graduate students.

Larger, multiperson, multi-institution grants may be more effective in building the institution but must be carefully managed, because HBCUs/MIs may sometimes feel that they are not participants in the decisions made by consortia and that their S&T capabilities are not adequately utilized by the prime contractors.

HBCU involvement in collaborative projects, when required, must be taken extremely seriously and, consequently, clearly and continuously documented by the prime contractors. It is necessary that HBCU selection as part of a research collective not be an afterthought and that ARL continuously assess the role of HBCUs/MIs in collaborative agreements. A defective model starts with the prime contractor (or group of prime contractors) organizing the proposal and then identifying the HBCU that fits the written proposal. This guarantees that little will be expected of the HBCU, discouraging any effort by either the prime contractor or ARL to mentor the HBCU participant and include it in significant and transparent ways. It is essential that the HBCUs/MIs be treated from the start as full partners in any collaborative project.

Faculty and administrators emphasized the benefits of securing funding for collaborative projects with colleagues at other research institutions; the by-products of collaborative dynamics were clearly identified, among them their role in adding fuel that further propels capacity building. HBCUs/MIs were open to receiving advice and mentoring and to collaborating with other institutions including HBCUs/MIs.

Senior administrators described poor communication between the institutions and the prime contractors. They would prefer an environment in which ARL and HBCU/MI leadership could discuss long-range research and development partnerships, including collaborative agreements and related funding. They suggest that the Army explore the possibility of using institutions that already have competitive and large research portfolios to serve as lead institutions and to mentor institutions that have not yet crossed that threshold. There are many lessons to be learned from the experience of faculty and leadership at institutions that have managed to become competitive.

While many contributors felt that ARL needs to play a more proactive role in introducing minority institutions to large cooperative programs that afford them meaningful participation, others emphasized the need for a balanced approach. Some researchers noted that the individual grant structure provides the best model for helping students find jobs. The concept of institution building was not a big concern for them; single-PI grants are seen as critically important because the funding comes directly to the PI, who can then support and manage the graduate student's research. Several suggestions were offered for improving individual grants (presumably those managed by ARO): alter the Broad Agency Announcement (BAA) to permit more travel and add a mandate to include undergraduates; focus on small contracts of 3 years duration with 2-year extensions if performance is good; and get the Army to help in finding internship opportunities for the students.

The faculty expressed a need for the ARL to help to secure more meaningful participation in the bigger cooperative programs and to give HBCUs/MIs a better forum to present their future research ideas. ARL was judged not to be proactive in helping to create collaborations with other universities and government labs.

NEED FOR SUSTAINED SUPPORT

Faculty and administrators were in accord that longer, sustained involvement with an ARL project has advantages over involvement typical of many single-investigator grants. While a 3-year grant can be an important part of a large, ongoing program, it is not enough by itself to develop a program. If not renewed, the resulting discontinuity in student support may be tantamount to forcing a new beginning. The general sense was that opportunities for contract and grant renewal could be integrated into the ARL program. In particular, single-investigator grants to early career faculty ought to be of sufficient duration and, where appropriate, include equipment, allowing these researchers to overcome the disadvantage that most HBCUs have of no start-up packages (or only a few) for early career faculty.

PIs worry about the inevitable end to the funding of a project and look for continued work with ARL. At the same time, of course, they are looking to find other source of support (e.g., NSF, NSA, and the National Oceanic and Atmospheric Administration [NOAA]) to complement ARL funding. Nonetheless the "3-5 years and out" profile generally does not allow the institution to build a useful foundation for itself or the Army.

Particular concern for continuous support of graduate students led to two suggestions:

1. ARL could adopt a model in which its support for Ph.D. students is expected to last 5 years (rather than 3) by way of grants/contracts, since such students must be supported for the duration of their Ph.D. work. Alternatively, 3-year grants could include 5 years of graduate student support; or, minimally, these grants could offer the possibility of no-cost extensions, so that students are not forced to jump into new projects in midstream.
2. Alternatively, and more in tune with the specific needs of minority graduate students, ARL might adopt a model that increases support for undergraduates or graduate U.S. minority students, particularly by means of supplements, as is done, for example, in the Diversity Supplement program of the National Institutes of Health (NIH), which supports graduate students and postdoctoral researchers. The NIH model, which has been used to support undergraduates as well,[1] might be useful to the Army.

This preference for sustained support did not overlook possible issues with longer term commitments. For example, during execution of a 3-5 (or more) year funded project, the ARL emphasis may shift. At a large institution, such changes in direction within an awarded project can be accommodated more easily because they have a broader talent pool and robust, in-place facilities. At small institutions (as is typical of many HBCUs), a change in direction may require new talent that is not readily available and/or facilities that are not in place, thereby derailing progress toward research students' degrees. Establishing HBCU/MI affiliations with collaborative entities such as University-Affiliated Research Centers (UARCs) may be a good way to help HBCUs/MIs weather such midproject redirection.

FACULTY AND STEM PROGRAM DEVELOPMENT, INCLUDING STUDENT DEVELOPMENT

Uniform enthusiastic support was expressed for the thesis that funds from ARL had been significant in assisting the development of local STEM programs. ARL funding has been used to carry out research that eventually led to new research centers or important research discoveries, or the strengthening of specific groups. These efforts have led to the establishment of new courses, applied Ph.D. programs, and the development of new curricula at the undergraduate and graduate levels.

[1] National Institutes of Health, "Research Supplements to Promote Diversity in Health-Related Research," last reviewed September 2, 2014, http://www.nigms.nih.gov/Research/Mechanisms/Pages/PromoteDiversity.aspx.

At each institution visited there were centers of excellence that ranged widely from one institution to the next in terms of technical area of emphasis, size, and numbers. When discussing these success areas, faculty and administrators acknowledged the ARL funding, and while recognizing that it might typically have provided 10-20 percent of the total research funding on which the center's development depended, they did give credit to ARL funding in many statements such as those paraphrased in the remainder of this section.

Faculty believe that the benefits ARL funding bring to their STEM programs are twofold. First such funding fosters interaction between graduate students and undergraduates. An informal communication network often develops in which students dialogue and contribute time to helping one another, thereby providing insight into the nature of the research being conducted. Second, the funds permit the faculty to consult with their ARL counterparts as well as colleagues at other universities. However, faculty also noted that the ARL dialogue is not frequent enough or robust enough. It was commonly observed that ARL could work with local faculty to develop an effective plan and process that increases the technical interaction between ARL researchers and university faculty and students.

In many cases, faculty have been innovative in blending graduate and undergraduate participation in their programs. They use various means, including providing elective research participation credits that can be earned. The underlying idea is to start the pipeline for research early. The program earns university credits.

Both administrators and faculty suggested that one of the goals of the HBCU/MI program at ARL is to build the institutional capabilities at funded universities. It is therefore critical to take a strategic view based on a full understanding of current strengths. This will help to develop a shared vision of the future with a plan for getting there. Administrators and faculty proposed that ARL increase funding to support the establishment of research centers, thereby enhancing the positive impact of ARL support on campus. Faculty and administrators appeared to be receptive to having ARL scientists help to plan and establish new programs of relevance to Army needs.

The faculty noted many examples of how initial ARL funding had led to further grants from DoD, NSF, and other agencies, which leveraged the initial research. Numerous examples were also provided where the faculty had been effective in forming strategic alliances with various companies to leverage their research. The fact that ARL has confidence in a university adds greatly to the school's credentials. However, the caliber of the research and of the students were the critical components of success.

EQUIPMENT AND FACILITIES

One may generalize about the lack of competitive facilities at HBCUs/MIs relative to larger research universities. Several discussants indicated that equipment for individuals (particularly those in start-up positions) and for multiple users in facilities was inferior to that at top-ranked competitors and collaborators and that the relatively small grants to HBCUs made equipment purchase and development difficult. In addition, they noted that in many instances, only minimal funding was available from university sources to cover maintenance and upgrades on equipment that had been painfully acquired. Important exceptions to this situation could be found in the centers of excellence that had grown over the years and that were often supported by multiyear, collaborative grants.

This lack of extensive infrastructure gives rise to a disadvantage in the competition for more equipment, a Catch-22 situation. For example, equipment proposals from the HBCU may often be designed to begin equipment development (new hires normally do not receive start-up packages; state funding yields buildings, rarely equipment), while the competing proposal from a large university will complement already existing infrastructure. Lack of technical assistance and minimal matching funds may also have an impact here. Furthermore, even if the equipment is successfully acquired, lack of continuing funding or related research funding may make the equipment at the HBCU/MI less effective after several years than it might have been at a major institution. All of these factors may lead proposal evaluators to see lower potential returns on investment in the HBCU/MI proposals.

One proposal for increasing access to facilities would have ARL establish and support regional centers where local universities, including HBCUs/MIs, would have access to core research equipment. (More likely, this would be a DoD effort rather than the effort of an individual Service laboratory.) Some attention was given to opportunities to acquire surplus ARL equipment as ARL continues to improve its own facilities. It was suggested that this might be done in a manner that would allow repurposing such surplus by targeted HBCUs/MIs. It was noted that civilian agencies such as the National Institute of Standards and Technology (NIST) already do this, but the HBCU/MI representatives believed that DoD does not.

ADMINISTRATION OF AWARDS

The process of initiating proposals and managing grant and/or contract funding can vary markedly from one institution to another. Those universities farther along on the path of research development will have strong central systems to support individual or group faculty efforts. By contrast, many HBCUs/MIs still find themselves in the situation whereby individuals must "go it alone" in dealing with the highly varied elements of award administration. Such procedures differ not only from grants and contracts at ARL but also of course, and markedly, from one funding agency to another. Administrative procedures vary widely from one school to the next, as can be exemplified by this brief sampling of comments from faculty and administrators:

- Grant processing is satisfactory; contract processing is more complex.
- There are problems with timing: Funds that arrive too late to hire students or postdoctoral researchers result in low expenditures during the early stages of programs and raise concerns for the sponsoring program manager at ARL.
- Many BAAs limit proposal submissions from a given university to two (sometimes only one) to minimize the load on sponsoring program managers. Internal competition at some schools may then choose the proposals that are best from the perspective of the local goals, eliminating from submission some that might ultimately be viewed best by the Army program manager. Perhaps in competitions targeted at HBCUs/MIs this limit on proposals might be waived or at least eased.
- DoD contracting and report requirements can be very different from those of other agencies. This is particularly true of 6.2 programs, which may take the form of contracts, not grants.
- There has been an issue with multiyear contracting. The faculty expressed the concern that ARL is restricting the HBCUs/MIs, unlike other universities, to yearly contracting.
- There is a difference in indirect costs between contracts and grants; perhaps the contract rate (which is lower) could be used. Also, no-cost extensions are not available.
- ARL could beneficially increase the pool of funds for which HBCUs/MIs may compete and the number of proposals that a given school is allowed to submit.
- In many instances the contracting is conducted and managed by the faculty. The faculty has no difficulty with the process, and funds appear to be released on time. There is a belief, however, that the BAA could beneficially be tailored for the university and that flexibility could be added to permit, or at least not to limit, travel as part of the effort.

STUDENT ENGAGEMENT WITH ARL

ARL has as one of its goals in the HBCU/MI program that students will become aware of, and to the extent possible, participate in Army programs. The funding and quality of research provides a relevance that the students appreciate. While the unstated goal of direct student involvement is eventual employment at ARL to help address issues of workforce diversity, this is neither anticipated nor measured by ARL. Faculty and students interviewed were well aware of the source of funding and, in most cases, of the relevance of the research to Army goals (or at least to the stated ARL research agenda). However, the student experiences varied widely from school to school and from one type of funding organization to another.

A major issue is the country of origin of the graduate student. At some of the schools visited, foreign students are in the majority, while in others they are in the minority. Access to the ARL campus may be quite difficult to achieve for those non-U.S. students, limiting their participation.

Another significant issue is geography. It is not surprising to find that visits by students and direct participation in ARL activities is easier for those close to the ARL, especially in view of highly limited travel budgets.

Some of the responses from different institutions are presented below, revealing the great diversity of student interaction with ARL:

- There is effective but not sufficient ARL funding to support a good fraction of the strong students from underrepresented groups at the undergraduate level.
- Summer support for undergraduates is not sufficient; 5-year support for Ph.D. students is essential. Projects provide support only for 3 years, and that is not the best funding model.
- Graduate students and postdoctoral researchers have had and continue to have positive interactions and engagements with ARL researchers over a range of research topics.
- Several undergraduate students have done summer internships at ARL facilities.
- Students supported by ARL grants may spend summers at the ARL facility as visitors. Of course, this is the time when faculty expect their students to be free from coursework and available for research in their own school-based laboratories. Except in the unusual case where the student's work at the ARL facility complements or supplements the funded program at the home institution, this summer visit to ARL is actually detrimental to the success of the funded program.
- There were two opposing opinions about the ARL student (and professor) intern opportunities:

  —Internships are a great experience, especially for those who would want to keep research connections with ARL. No one knew, however, if any student intern had ever been hired by ARL.

  —Summer internships, while valuable, probably disrupt student progress toward completing a project task, thesis, or dissertation and, therefore, may add time to degree attainment. Discussion followed on the value of connecting the thesis or dissertation research to internship tasks at the beginning of a funded project so that no time is lost. This would take some attentive planning on the part of the ARL program manager and the funded institution.
- Small grant sizes limit opportunities for associated student travel and internship at ARL.

ARL–UNIVERSITY INTERACTIONS

Among the ARL goals for its HBCU/MI program is increased interaction between ARL researchers and faculty and students at the university. This interaction can take many forms. The closest might be cooperation in establishing program goals, followed by direct collaboration, at the ARL site or by the back-and-forth exchange of data, specimens to characterize, or other research items. An intermediate level of collaboration might entail regular communication during performance of the grant/contract, while minimal interaction might consist of infrequent communication that simply makes university personnel aware of the intended Army goals in sponsoring the stated research.

Administrators and faculty frequently indicated their desire for closer relationships with ARL, and they expressed some frustration that this was not happening. The following comments are particularly relevant to this issue:

- The dynamics for early career faculty often involve senior faculty who know the right contact(s) at ARO. A recently funded junior faculty member explained that the relevance of his research to the Army was not clear to him until a senior faculty member had identified for him an Army white paper that fit well with his (the junior faculty member's) research. Well-connected senior faculty are able to find opportunities and mentors before and during the award process. Many suggestions were made for how the process could be modified if the Army wants an HBCU/MI to have a serious capability to address its needs and goals.
- ARL often fails to provide sufficiently detailed and timely reviews of research white papers submitted by HBCU/MI researchers. Specific comments on the quality of the proposed science would be welcomed by local researchers. More feedback from ARL program managers on failed white papers and proposals would help faculty target for success. Developing successful proposals depends not only on technical content but also on format, which varies from agency to agency. There may be value in ARO's developing and offering a primer on how to write white papers, perhaps by Webinar.
- Several faculty noted poor feedback on failed proposals. While reviewer comments may have been shared, there did not seem to be opportunities to discuss them with program managers. It is not clear where the responsibility for this shortfall may lie. Faculty also indicated a desire for more interaction with the program managers during program activity.
- Many faculty members expressed the desire for greater participation in the preparation of BAAs prior to their issuance. This would imply a very different relationship with ARL in which strategic program goals are mutually developed and shared.
- ARL encourages participation of HBCU/MI students and faculty in Army-related projects by making opportunities available at the ARL facilities. While this is a desirable goal and may lead to valuable experiences as well as have an eventual impact on ARL hiring, there are aspects of current programs that could be handled better. Faculty are rarely involved directly at ARL facilities, and it is difficult for them to link to on-going research programs. Of course, in addition to the fact that ARL program managers at ARO are not only geographically distant from most funded sites, they do not work with the principal investigators by the very nature of their jobs. Meanwhile, faculty members are looking for closer interactions during the performance of their funded programs.
- Many of the comments by faculty and administrators expressed the desirability of ARL involving the HBCUs/MIs in its higher level, strategic planning in order to build capability of value to the Army. ARL might select one or more HBCUs/MIs as an affiliate laboratory. This same idea was floated with a focus on creation of a core facility of value to other schools and to ARL.
- Some suggested that in addition to individuals seeking relationships with ARL staff and program managers, it would be beneficial to have ARL seek them out. For example, ARL could be invited to attend the annual reviews of research that are hosted by some of the HBCUs/MIs.
- HBCUs/MIs could use the mentoring of ARL to find others in the Army that could fund continued research work related to the original ARL funding. It appeared also that single principal investigators could get better mentorship in connecting to other, non-ARO divisions of ARL for follow-on research support. A related discussion focused on assisting HBCUs/MIs to achieve access to and time on Army/DoD major shared resources (e.g., supercomputing centers).
- Some faculty participants referred to past programs (presumably no longer available) wherein ARL would help a university beyond a specific R&D project. For example, ARL had sent scientists to campuses to help establish new programs at universities. ARL would send its employees to a university to work on theses at the sponsored university. ARL would hire sponsored university professors to teach a technical topic at Army facilities nationally/globally. The discussants saw such non-project-specific initiatives as extremely valuable to STEM capability development and to the reputation of an HBCU/MI as well as clearly supportive of the longer-term interests of ARL and the Army.
- Some participants at universities located far from ARL raised an important issue that they suggested needs further attention. Geographic collocation plays a significant role in collaborations

between many universities and nearby government research laboratories in many agencies. The Army could identify a model or models that make it possible for scientists and researchers to participate collaboratively at research facilities that are not geographically convenient. They asked: How can the Army overcome the limitations of geography so that researchers can collaborate at any ARL facility?

FUNDING CONTEXT

HBCU/MI faculty share an issue similar to that encountered by faculty at all major research institutions—namely, that breaking into the club of those funded by a given ARL program manager is more difficult than getting continuing funding once a track record has been established. This is the underlying raison d'etre for early-career-investigator grants. There are several elements at play here: lack of extensive infrastructure, the reputation of the institution, and length and continuity of funding.

PIs carry with them the cachet of the institution they represent. Program managers at ARO not only sit and wait for proposals; they seek out potential fits to their program areas to address perceived needs, as do program managers at ARL directorates. If they do not look to an HBCU/MI or are unaware of potential opportunities there, it puts potential PIs at a disadvantage. Some faculty seemed to think that this issue needed to be resolved by ARL action; others felt that it was up to the university faculty to bring their skill set to the attention of ARL individual and/or collective action. Probably both efforts are required.

B

Summary Description of Army Research Laboratory Collaborative Research Programs[1]

Currently, the Army Research Laboratory (ARL) outsources 80 percent of its research program to academia, with over 250 academic partners in all 50 states, and to industry, through a mix of grants, cooperative agreements, contracts, and other forms of transaction.[1]

PARTNERSHIP IN TRANSITION PROGRAM

Under a special initiative of the Assistant Secretary of the Army for Acquisition, Logistics, and Technology (ASA[ALT]), the Partnership In Transition Program (PIRT) is an Army–HBCU (historically black college or university) program established as the second phase of the former Battlefield Capability Enhancement (BCE) program. It employs a center of excellence (COE) model. The purposes of the program are as follows: (1) to enhance the research programs and capabilities of a select number of high-interest scientific and engineering disciplines through Army-relevant, topic-focused, near-transition-ready innovative research and (2) to strengthen the capacity of HBCUs to provide excellence in education and provide opportunity for them to conduct research critical to the national security functions of the Department of Defense (DoD). The PIRT Broad Agency Announcement was released in July 2010 and closed October 2010. There were 11 topics and numerous proposals per topic; it was highly competitive. Five proposals were selected and awarded to four HBCU-led consortia in four topic areas. Each cooperative agreement has a manager and a comanager, one from ARL/ARO and the other from ARL Directorates (or the Engineer Research and Development Center [ERDC]) to enable transition. Programs are funded at between $400,000 and $500,000 per year for 3 to 5 years.

COLLABORATIVE RESEARCH ALLIANCES

Collaborative research alliances (CRAs) are alliances that bring together expertise from government, academia, and industry to address some of the fundamental scientific and technological underpinnings of our military defense systems. ARL has a history of successful collaborations, bringing together strong research talent from government, academia, and industry to develop creative and novel capabilities for the Army. Each CRA possesses unique strengths, and its developments will be transitioned into and aligned with the relevant needs and directions of the ARL enterprise. This ARL enterprise has the unique focus to deliver enabling capabilities that integrate state-of-the-art techniques, knowledge, and experience. Through the collaborations among the many participants, the enterprise is expected to move the ARL forward, along with its major laboratory programs, toward a vision of robust

[1] This appendix summarizes descriptions of ARL partnership opportunities described on the ARL website at http://www.arl.army.mil/www/default.cfm?page=9, accessed May 25, 2014.

multiscale control over Army materials. Typically a university will lead this effort. Duration and level of funding vary with the programs.

COLLABORATIVE TECHNOLOGY ALLIANCES

The collaborative technology alliances (CTAs) are partnerships between Army laboratories and centers, private industry, and academia that are focusing on the rapid transition of innovative technologies to the warfighter to enable the Army's Future Force. The collaboration between industry, academia, and government is a key element of the CTA concept, as each alliance member brings with it a distinctly different approach to research. Academia is known for its cutting-edge innovation; the industrial partners are able to leverage existing research results for transition and to deal with technology bottlenecks; the Army Research Laboratory's researchers keep the program oriented toward solving complex Army technology problems. Thus, multidisciplinary research teams are generating the complex technology needed to solve the Army's complex problems. This approach enables the CTA program to bring together world-class research and development talent and focus it on Army-specific technology objectives for application to Army needs. Typically, industry will lead these efforts. Programs are typically funded at $5 million to $8 million per year for 5 to 8 years.

INTERNATIONAL COLLABORATIVE ACTIVITIES

ARL makes use of international collaborative activities with allied defense establishments to leverage its mission-funded R&D investments. ARL has a vigorous international program with the lead on numerous active and proposed bilateral agreements, as well as support to other Department of the Army (DA) and DoD agencies in their cooperative programs. These agreements enable cooperative research programs with allies in selected technology areas where their strengths complement ARL's and offer good leveraging opportunities. For example, tactical information processing with Germany, solid state laser research with Israel, laser beam attenuation with Israel, and fuel cell research with Singapore. ARL has established a pioneering cooperative research program with the United Kingdom in network and information science through an International Technology Alliance (NIS-ITA). Under this arrangement, the United States and the United Kingdom jointly created and funded a consortium of industry and academic partners to perform research in which both countries, and the consortium partners, benefit and share intellectual property rights and will commence work to transition projects beyond the fundamental research program. In addition, extensive cooperative activity takes place through multinational forums including The Technical Coordinating Panel (TTCP), the North Atlantic Treaty Organization (NATO) Research and Technology Organization (RTO), and the five country senior national representative-Army (SNR(A)) working groups. Duration and level of funding vary with the programs.

UNIVERSITY-AFFILIATED RESEARCH CENTERS (UARCs)

A university-affiliated research center (UARC) is a strategic DoD research program that is associated with a university. UARCs were established to ensure that essential engineering and technology capabilities of particular importance to the DoD are maintained. UARCs are designed to provide critical mass in research areas that meet Army and DoD future needs and anticipated combat requirements. They are university-led collaborations between universities, industry, and Army laboratories that conduct basic, applied, and technology demonstration research. The universities, considered to be at the forefront of science and innovation in any given research area, provide dedicated facilities and share space with Army and industrial participants. The industrial partners provide competence in related technologies and expertise in transitioning technologies from the lab to the market, and they share the costs. Each UARC conducts research where breakthroughs are likely to enable revolutionary capabilities for our warfighters. UARCs are typically funded at between $5 million and $10 million per year for 5 years (renewable). To date no HBCU/MI has been a UARC or a formalized partner with a designated UARC.

CENTERS OF EXCELLENCE

The Army COE program for HBCUs/MIs was established in FY2004 at the initiative of ASA (ALT). Its purpose was to explore and mature technologies with potential to enhance the Future Force Team with battle labs for requirements input and continuing refinement/prioritization. It also was to intended to enable and enhance HBCU/MI organic research capabilities. The COEs comprise university-led focused initiatives and competitive contracts. FY2008 and FY2009 were the last funding years for this program.

MULTIDISCIPLINARY UNIVERSITY RESEARCH INITIATIVE

This program is part of the University Research Initiative (URI) and is managed by the DoD research offices: the Office of Naval Research (ONR), the ARO, and the Air Force Office of Scientific Research (AFOSR). Awards take the form of grants. The Multidisciplinary University Research Initiative (MURI) program supports university teams whose research efforts intersect more than one traditional science and engineering discipline. MURIs are typically funded at approximately $1.25 million per year for 3-year periods.

C

Biographical Sketches of Members of the Committee

WESLEY L. HARRIS, *Chair*, NAE, is the Charles Stark Draper Professor of Aeronautics and Astronautics and director of the Lean Sustainment Initiative at the Massachusetts Institute of Technology. He was elected to the NAE "for contributions to understanding of helicopter rotor noise, for encouragement of minorities in engineering, and for service to the aeronautical industry." He has performed research and published in refereed journals in the following areas: fluid mechanics; aerodynamics; unsteady, nonlinear aerodynamics; acoustics; lean manufacturing processes; military logistics and sustainment, and hemodynamics. Dr. Harris has substantial experience as a leader in higher education administration and management. He also has demonstrated outstanding leadership in managing major national and international aeronautical and aviation programs and personnel in the executive branch of the federal government. He is an elected fellow of the Aerospace Industries Association of America, Inc., the American Helicopter Society, and the National Technical Association for personal engineering achievements, engineering education, management, and advancing cultural diversity. He earned a Ph.D. and an M.S. in aerospace and mechanical sciences at Princeton University and a B.S. in aerospace engineering (honors) from the University of Virginia.

SANDRA BEGAY-CAMPBELL is a principal member of the technical staff at Sandia National Laboratories. Ms. Begay-Campbell leads Sandia's technical efforts in the Renewable Energy Program to assist tribes with renewable energy development. She served as a member of the NSF's Committee on Equal Opportunities in Science and Engineering. As a member of the Navajo Nation, Ms. Begay-Campbell's perspective incorporates her cultural values into a technical environment. Ms. Begay-Campbell is the former executive director of the American Indian Science and Engineering Society (AISES), a nonprofit organization whose mission is to increase the number of American Indian scientists and engineers. She subsequently worked at Los Alamos National Laboratory before joining Sandia. She has served on two committees for the National Academy of Engineering: the Committee on Diversity of the Engineering Workforce and the Committee on Engineering Studies at Tribal Colleges. Ms. Begay-Campbell received a B.S. in civil engineering from the University of New Mexico and an M.S. in structural engineering from Stanford University.

FRANK CAPPUCCIO is the President and CEO of Cappuccio and Associates LLC. He recently retired from Lockheed Martin Corporation as executive vice president and general manager of its famed Skunk Works, tasked with the pursuit, capture and selective execution of new business for Lockheed Martin Aeronautics Company. Prior to that, Mr. Cappuccio was the Lockheed Martin corporate vice president of the Joint Strike Fighter Program. He also served as vice president for programs and technology for the company's Aeronautics Sector in Bethesda, Maryland. He has over 45 years of comprehensive and diverse management and engineering experience in acquisition, development, and deployment of high-tech products ranging from navigational computers to missiles and tactical fighters. Mr. Cappuccio holds an M.B.A. from Adelphi University, an M.S. in mechanical engineering from Columbia University and a B.S. in mechanical engineering from City College of New York.

CARLOS CASTILLO-CHAVEZ is a regents professor and a Joaquin Bustoz Jr. Professor of Mathematical Biology at Arizona State University and the founding director of the Simon A. Levin Mathematical, Computational and Modeling Sciences Center and has coauthored over 200 publications at the interface of the life, social, and mathematical sciences. He has had 33 Ph.D. students, a group that includes 19 individuals from underrepresented groups. Recognitions of his work include three White House Awards (1992, 1997, and 2011), the 2010 American Mathematical Society Distinguished Public Service Award, and the 2007 AAAS Mentor award. He is a fellow of the American Association for the Advancement of Science (AAAS), the Society for Industrial and Applied Mathematics (SIAM), the American Mathematical Society (AMS), and the American College of Epidemiology (ACE). He has held honorary professorships from Xi'an Jiatong University in China and Universidad de Belgrano in Argentina. He was appointed a Stanislaw M. Ulam Distinguished Scholar at Los Alamos National Laboratory, a Cátedra Patrimonial at UNAM in Mexico, and a Martin Luther King, Jr., Professor at MIT. He is a member of the Board of Higher Education at the National Academy of Sciences (2009-2016) and serves on President Barack Obama's Committee on the National Medal of Science.

VADM PAUL G. GAFFNEY II, NAE, served as the seventh president of Monmouth University from 2003 to 2013; he is president emeritus and currently part of Monmouth's Urban Coast Institute as its first fellow. A retired Navy vice admiral, he was president of the National Defense University from 2000 to 2003. Prior to that, he was the Chief of Naval Research with responsibility for Department of the Navy science and technology investment. He was appointed to the statutory U.S. Ocean Policy Commission and served during its full tenure from 2001 to 2004. In his military career he headed the Navy's worldwide operational meteorology and oceanography program and commanded the Naval Research Laboratory. He has been recognized with a number of military decorations: the Naval War College's J. William Middendorf Prize for Strategic Research, the Outstanding Public Service Award from the Virginia Research and Technology Consortium, and the Potomac Institute's Navigator Award. He is a fellow of the American Meteorological Society, has served on several boards of higher education, was a member of the National Research Council's Ocean Studies Board and chaired the Federal Ocean Research Advisory Panel. He is a director of Diamond Offshore Drilling, Inc. Admiral Gaffney is a graduate of the U.S. Naval Academy and holds an M.S. in ocean engineering from Catholic University. He graduated from the Naval War College with highest distinction. He earned an M.B.A. from Jacksonville University. The University of South Carolina, Jacksonville University, and Catholic University have awarded him honorary doctorates.

MICHAEL T. NETTLES is senior vice president and Edmund W. Gordon Chair at the Educational Testing Service (ETS). Dr. Nettles' research covers a broad spectrum of educational policy topics, including educational assessment at the elementary, secondary and postsecondary levels; student achievement; educational opportunity; access and equity; faculty compensation and rewards; and financing higher education. Dr. Nettles is a member of the Bank Street College of Education Board of Trustees and the National Center for the Improvement of Educational Assessment. He also serves on the board of the National Science Foundation-sponsored Center on Research on Teaching and Learning and the board of the Center for Enrollment Research, Policy and Practice at the University of Southern California. Dr. Nettles was initially at ETS from 1984 to 1989, first as a research scientist and later as a senior research scientist. From 1989 to 2003, he served as vice president for assessment for the University of Tennessee system and as a professor of education for 12 years at the University of Michigan. Prior to returning to ETS in 2003, he served for a decade on the National Assessment Governing Board, which oversees and develops policies for the National Assessment of Educational Progress; for 8 years on the board of trustees of the College Board, which owns the SAT, the Advanced Placement course and exams, and other educational products and services; and for 4 years on the Graduate Record Examination board. Dr. Nettles earned his B.S. in political science at the University of Tennessee and two master's degrees, one in political science and the other in higher education, and a Ph.D. in higher education from Iowa State University.

LYLE H. SCHWARTZ, NAE, is retired director, Air Force Office of Scientific Research. He was professor of materials science and engineering at Northwestern University for 20 years and director of Northwestern's Materials Research Center for 5 of those years. He then became director of the Materials Science and Engineering Laboratory at the National Institute of Standards and Technology, where he served for more than 12 years. His experience there included metals, ceramics, polymers, magnetic materials, techniques for characterization, and standardization of these characterization techniques, and his responsibilities included management of the R&D agenda in the context of a government laboratory. Dr. Schwartz subsequently assumed responsibility for basic research on structural materials of interest to the U.S. Air Force in addition to the areas of propulsion, aeromechanics, and aerodynamics. He completed his government service as director of the Air Force Office of Scientific Research with responsibility for the basic research program of the Air Force. His current interests include government policy for R&D, particularly for materials R&D, STEM education at K-12 levels, and enhanced public understanding of the roles and importance of technology in society. He is a member of the National Academy of Engineering. Dr. Schwartz received both a B.S. in engineering and a Ph.D. in materials science from Northwestern University.

THEDA SKOCPOL, NAS, is the Victor S. Thomas Professor of Government and Sociology at Harvard University. At Harvard, she served as dean of the Graduate School of Arts and Sciences (2005-2007) and as director of the Center for American Political Studies (2000-2006). In 1996, Dr. Skocpol served as president of the Social Science History Association, an interdisciplinary professional group, and in 2002-2003, she served as president of the American Political Science Association during the centennial of this leading professional body. Dr. Skocpol's research focuses on U.S. social policy and civic engagement in American democracy, including changes since the 1960s. She has recently launched new projects on the transformations of U.S. federal policies in the Obama era. Her books and articles have won numerous awards, including the 1993 Woodrow Wilson Award of the American Political Science Association for the best book in political science for the previous year. In 2007, she was awarded the Johan Skytte Prize in Political Science. Dr. Skocpol also belongs to the American Academy of Arts and Sciences (elected 1994), the American Philosophical Society (elected 2006), and the National Academy of Sciences (elected 2008). She earned an M.A. and a Ph.D. in sociology from Harvard University and a B.A. in sociology from Michigan State University.

D

Acronyms

AFOSR	Air Force Office of Scientific Research
AANAPISI	Asian-American and Native American Pacific Islander-Serving Institution
AIHEC	American Indian Higher Education Consortium
ARL	Army Research Laboratory
ARO	Army Research Office
BAA	Broad Agency Announcement
BCE	battlefield capability enhancement
BD	bridge to the doctorate
CGS	Council of Graduate Schools
COE	center of excellence
CRA	Collaborative Research Alliance
CTA	Collaborative Technology Alliance
DoD	Department of Defense
HACU	Hispanic Association of Colleges and Universities
HBCU	historically black colleges and universities
HSI	Hispanic-serving institution
ITA	International Technology Alliance
LSAMP	Louis Stokes Alliances for Minority Participation
MI	minority institution
MURI	Multidisciplinary University Research Initiative
NATO	North Atlantic Treaty Organization
NIH	National Institutes of Health
NRC	National Research Council
NSF	National Science Foundation
OSD	Office of the Secretary of Defense
ONR	Office of Naval Research
PIRT	Partnership in Research Transition
R&D	research and development
RFP	request for proposals

RTO	research and technology organization
S&T	science and technology
STEM	science, technology, engineering, and mathematics
TCU	tribal colleges and universities
UARC	university-affiliated research center
URM	underrepresented minority